Construction Economics in the Single European Market

Construction Economics in the Single European Market

Edited by

B. Drake

for

The Committee of European Construction Economists

Taylor & Francis
Taylor & Francis Group

LONDON AND NEW YORK

Published 1995 by Taylor & Francis
2 Park Square, Milton Park, Abingdon, Oxon, OX14 4RN
52 Vanderbilt Avenue, New York, NY 10017, USA

First issued in paperback 2020

Taylor & Francis is an imprint of the Taylor & Francis Group, an informa business

First edition 1995

© 1995 Comité Européen des Economistes de la Construction

A catalogue record for this book is available from the British Library

Publisher's Note
The publisher has gone to great lengths to ensure the quality of this reprint but points out that some imperfections in the original may be apparent

ISBN 13: 978-0-367-57975-3 (pbk)
ISBN 13: 978-0-419-18980-0 (hbk)

Contents

Contents

Comité Européen des Economistes de la Construction Construction Economics European Committee

Annuaire Directory of members

Siege Social/ Registered Office:	8 Avenue Percier 75008 Paris FRANCE Tel: 1–45 63 30 41
Secretariat:	Miss Marianne Tissier 12 Great George Street Parliament Square London SW1P 3AD Tel: 71–222 7000 Fax: 71–334 3790

Organisations Membres/Member Societies

Belgium	Union Belge des Géomètres-Experts Immobiliers Rue du Nord 76 1000 Brussels Tel: 32 2 218 07 13 Fax: 32 3 219 31 47

Denmark	The Danish Association of Construction Economists Danske Byggeokonomer Kronprinsessegade 7 DK-1306 Copenhagen K Tel: 45 33 33 05 90 Fax: 45 33 33 05 91
Finland	The Finnish Value and Cost Engineering Association (VACE) Mrs Tuula Laitinen, Secretary Kurtinmäentie 12 02780 Espoo Tel: 358 0 810 758 Fax: 358 0 810 470
France	Union Nationale des Techniciens Economistes de la Construction 8 Avenue Percier 75008 Paris Tel: 1–45 63 30 41 Fax: 1–42 56 14 52 M. Martial Monbeig-Andrieu
Ireland	Society of Chartered Surveyors in the Republic of Ireland 5 Wilton Place Dublin 2 Tel: 353–1–676 5500 Fax: 353–1–676 1412 Mr Tony Smith
Netherlands	Nederlandse Vereniging Van Bouwkostendeskundigen Prins Mauritslaan 29–39 Postbus 302 1170 AH BADHOEVEDORP Netherlands Tel: (31) 20 658 0250 Fax: (31) 20 659 2477
Portugal	Sindicato dos Agentes Technicos de Arquitectura e Engenharia Calcado do Combro 127-2 Dt 1200 LISBOA PORTUGAL

Spain	Consejo General de la Arquitectura Técnica de España Paseo de la Castellana 155, 1 28046 Madrid SPAIN Tel: (91) 570 15 35 Fax: (91) 571 28 42
Switzerland	Swiss Professional Association for Construction Economics *Headquarters* Habsburgerstr. 30 6003 Lucerne Tel: 041 23 43 77 Fax: 041 23 43 80 *Address for correspondence* Hurdenerstr. 117 8640 Hurden
United Kingdom	Royal Institution of Chartered Surveyors 12 Great George Street Parliament Square London SW1P 3AD Tel: 71–222 7000 Fax: 71–222 9430

Comité Européen des Economistes de la Construction
Construction Economics European Committee

Liste des membres des delegations et commissions
List of members of delegations and commissions

Member of UK Delegation	Professor Brian Atkin Dept of Construction Management University of Reading Whiteknights Reading RG6 2BU UNITED KINGDOM Tel: 0734–875123 Fax: 0734–313856
Observer Italy	Mr Claudio Caresia Via Vergani Marelli 5 20146 Milan ITALY
Past President (1988–91) Member of Spanish Delegation National Co-ordinator	Sr Antonio Castrillo Canda Consejo General de la Arquitectura Técnica de España Paseo de la Castellana 155, 1 28046 Madrid SPAIN Tel: (91) 570 15 35 Fax: (91) 571 28 42

Member of Dutch Delegation	Mr W. Cornelissen Bouwcentrum Advies BV Weena 760 Postbus 299 3000 AG Rotterdam THE NETHERLANDS Tel: 31–10 430 92 04 Fax: 31–10 412 11 15
Individual member representing Italy	Professor Guido Dandri Facolta di Architettura Universita degli Studi di Genova Stradone di S. Agostino n. 37 16123 Genova ITALY Tel: 353/8859
Honorary Secretary Member of Belgium Delegation National Co-ordinator	Mr Christian Deprez Widnell Europe SA Rue Blanche 15 Boîte 3 1050 Brussels BELGIUM Tel: 2–538 92 02 Fax: 2–539 04 98
Member of Portuguese Delegation	Mr Jack Francisco 8 Horwood Close Headington OXFORD OX3 TRF
Member of Belgian Delegation	Mr Jan de Graeve 5 Avenue de Meysse 1020 Brussels Tel: 2–268 10 25 Fax: 2–262 10 33
Member of UK Delegation National Co-ordinator	Mr John Gregory Gleeds Wilford House 1 Clifton Lane Wilford Nottingham NG11 7AT Tel: 0602 455233 Fax: 0602 455159

Member of Portuguese Delegation	Mr Victor Goncalves
Member of Finnish Delegation National Co-ordinator	Mr Kauko Nygren Chairman, VACE HKR Kasarminkatu 21 00130 Helsinki Tel: 358 0 166 2511 Fax: 358 0 625 940
Member of Danish Delegation National Co-ordinator	Mr Michael Hartmann The Danish Association of Construction Economists Danske Byggeokonomer Kronprinsessegade 7 DK-1306 Copenhagen K DENMARK Tel: 33 33 05 90 Fax: 33 33 05 91
Member of Danish Delegation	Adam Trier Jacobsen The Danish Association of Construction Economists Danske Byggeokonomer Kronprinsessegade 7 DK-1309 Copenhagen K DENMARK Tel: 33 33 05 90 Fax: 33 33 05 91
Member of Danish Delegation	Mr Nikolaj Jensen The Danish Association of Construction Economists Danske Byggeokonomer Kronprinsessegade 7 DK-1306 Copenhagen K DENMARK Tel: 33 33 05 90 Fax: 33 33 05 91

Member of Irish Delegation	Mr David Kelly Healy Kelly & Partners Duncairn House 14 Carysfort Avenue Blackrock, Co Dublin IRELAND Tel: 353–1–283 1116 Fax: 353–1–288 2481
Member of Portuguese Delegation National Co-ordinator	Mr Jose M. O. Lirio de Carvalho Sindicato dos Agentes Tecnicos de Arquitectura e Engenharia Calcada do Combro 127-2 Dt 1200 Lisboa PORTUGAL Tel: 351 1 342 1597/8416 Fax: 351 1 343 9859
Member of Irish Delegation	Mr D. Laurence Martin D.L. Martin and Partners 9 Mount Crescent Dublin 2 EIRE Tel: 353–1–661 4873 Fax: 353–1–661 0336
Member of Dutch Delegation	Mr Willem Meijer Staalstraat 4 5344 KG Oss NETHERLANDS Tel: 4120 24071 Fax: 4120 47875
Senior Vice President Member of French Delegation National Co-ordinator	Mr Gérard Méganck 48 rue des Petites Ecuries 75010 Paris FRANCE Tel: 1 42 47 19 08 Fax: 1 48 24 28 65 OR c/o UNTEC

Member of French Delegation	Mr Martial Monbeig-Andrieu UNTEC 8 Avenue Percier 75008 Paris FRANCE Tel: 1 45 63 30 41 Fax: 1 42 56 14 52
Member of French Delegation	Mr Jacques Moreau c/o UNTEC 8 Avenue Percier 75008 PARIS France Tel: 1–45 63 30 41 Fax: 1–42 56 14 52
Member of Spanish Delegation	Mr Jose Nagueire c/o Consejo General de la Arquitectura Tecnica de España Paseo de la Castellana, 155 28046 Madrid SPAIN Tel: (91) 570 15 35 Fax: (91) 571 28 42
Member of French Delegation	Pierre Mit UNTEC 8 Avenue Percier 75008 Paris FRANCE
Individual member representing Romania	Crina Oltean – Dumbrava Dept of Const. Managt. & Tech. Tech. Uni. of Cluj-Napoca Fac. of Civil Engineering 15 C Daicoviciu St. 3400 Cluj-Napoca Romania Tel/Fax: 40 64 192 055
Member of Finnish Delegation	Ari Pennanen The Finnish Value and Cost Engineering Association (VACE) Tinasepäntie 45 00620 Helsinki FINLAND Tel: 0–757 16 61 Fax: 0–792 635

Member of French Delegation	Mr Louis Plagnol 7 rue Cite-Foulc 3000 Nîmes FRANCE Tel: 66 67 83 01 Fax: 66 76 06 92
Member of Spanish Delegation	Sr A Ramirez de Arellano Consejo General de la Arquitectura Tecnica de España Paseo de la Castellana, 155 28046 Madrid SPAIN Tel: (91) 570 15 35 Fax: (91) 571 28 42
Member of the UK Delegation	Mr Douglas Robertson 85–87 Clarence Street Kingston Upon Thames Surrey KT1 1RB UNITED KINGDOM Tel: 81–549 0102 Fax: 81–547 1238
Member of Dutch Delegation	Mr Ko L. Staneke Staneke Bouwkostenmanagement Lagedijk 166 1544 BL Zaandijk NETHERLANDS Tel: 75–21 21 61 Fax: 75–21 85 23
Observer Switzerland	Mr Felix Trefzer Geschaftsfuhrer Schweizerische Zentralstelle fur Baurationalisierung Zentralstrasse 153 8040 ZURICH Switzerland Tel: 1–451 2288 Fax: 1–451 1521

Member of Belgian Delegation	Mr Wachtelaer c/o C. Deprez Esq Widnell Europe SA Rue Blanche 15 Boîte 3 1050 Brussels BELGIUM Tel: 2–538 92 01 Fax: 2–539 04 98
President	Mr Rob de Wildt RIGO Research en Advies BV De Ruyterkade 139 Postbus 2805 1000 CV Amsterdam Tel: 31–20 522 11 22 Fax: 31–20 627 68 40
Junior Vice President Member of Irish Delegation National Co-ordinator	Mr Michael J. T. Webb Patterson Kempster & Shortall 24 Lower Hatch Street Dublin 2 EIRE Tel: 353–1–676 3671 Fax: 353–1–676 3672
Member of Swiss Delegation National Co-ordinator	Mr Martin Wright PBK Projektmanagement, Bauadministration und Kostenplanung Hurdenstrasse 117 8640 Hurden SWITZERLAND Tel: 55/486363 Fax: 55/486372
Observer Austria	Mr Josef Mahlknecht Marktgasse 7a 6800 Feldkirch AUSTRIA

Contributors

Brian Atkin
CEEC, UK

Tony Burton
Gardiner & Theobald, UK

Bent Christensen
CECC, Denmark

John Connaughton
Davis, Langdon & Everest, UK

Brian Drake
President, CEEC

Michael C. Hartmann
Senior Vice President, CEEC

Keith Hudson
BCIS, UK

Robert Martin
Commission of the European Communities

Jacques Moreau
CEEC, France

Gerry O'Loughlin
CEEC, UK

David Owen
Chesterton Soprec Blumenauer, UK

John Pelling
AEEBC, UK

Ari Pennanen
CEEC, UK

Antonio Ramirez
CEEC, Spain

Douglas Robertson
CEEC, UK

Bjørn Roepstorff
Risk Management Consult ApS, Denmark

C.F. Stoker
Gardiner & Theobald Engineering Services, UK

Martin Wade
CEEC, UK

Michael Webb
CEC, UK

Rob de Wildt
CEEC, Netherlands

Preface

This book comprises papers presented at the joint CEEC/UNTEC conference on Construction Economics, held at the World Trade Centre, La Defense, Paris in May 1993.

CEEC is the Committee of European Construction Economists; formed in 1979, the Committee now comprises representatives from some 12 European countries. UNTEC is the French organization of construction economists: the Union Nationale des Economists de la Construction.

CEEC is most grateful to UNTEC for undertaking the physical organization of the conference and for underwriting the financial arrangements.

Many of the papers in this volume derive from the substantial programme of studies that CEEC has in hand. These papers, produced by busy practitioners to whom I express my gratitude, make no claim to academic rigour. They are rather a record of certain aspects of construction economics practice at the dawn of the Single European Market. In preparing these papers for publication I have therefore not sought to make extensive amendment or even to correct some highly original English: better, I felt, to preserve the charm and feel of the original.

Contrary to ill-informed opinion in the UK, there are many people practising construction economics in continental Europe. To have acted as their President for three years has been a rewarding, enlightening and enjoyable privilege. I believe a similar experience awaits the readers of this book.

Brian Drake

1

Introduction

Brian Drake

I have two tasks: to explain the aims of CEEC, and to explain the aims of the conference on which this book is based.

1.1 PURPOSE OF CEEC

Why was CEEC formed, on an UNTEC initiative, in 1979 and what purpose does it serve?

Our Statutes have a number of quite elevated, though sensible, objectives talking about exchanging experience and information, promoting training and harmonizing standards, establishing guidelines for practice, harmonizing legislation affecting the profession, coordinating working methods and establishing European statistics.

1.1.1 Defensive measures

But I believe that the real motivation for the formation of CEEC lay in the statutes, which read:

> ensuring adequate representation of qualified persons who are responsible for construction economics in the European Commission and other European institutions.

and

studying existing and proposed legislation and regulations relating to construction economics...

In other words, it was a defensive measure against the effects of Brussels decisions on a profession of which, in those days, the Commission knew little.

This was, and is, a very reasonable safeguard; in the kind of horsetrading that necessarily goes on in the EC or in GATT negotiations it is very easy for the interests of smaller groups to be sacrificed for the sake of large and noisy groups able to deliver votes; yes, I *am* referring to agriculture.

The years since 1979 have seen the formation of many similar single-interest groups on a pan-European basis, and of course for very similar reasons. CEEC remains well aware of the central importance of this part of its role, and many of its efforts are directed towards relations with the Commission. In return, we are now frequently consulted on matters relevant to our profession and our interests, and the value of our contributions has been acknowledged by the Commission: most recently, work by Rob de Wilt and colleagues on liability in construction has been commended by Commissioner Martin Bangemann.

1.1.2 Work programme

But a defensive posture would be an uncomfortable one for most of us, so we also have quite a substantial work programme. In fact, the papers that were given at the conference on which this book is based were largely the reports of the individuals responsible for various items under the work programme. But in addition to these there is important work in hand by Gerard Meganck on the special problems that may have been created for UNTEC members by the terms of the mutual recognition directive; papers have been submitted at the request of the Commission on obstacles to practice within and outside the EC for use in GATT negotiations; the liability question has been very demanding on our members; contributions have been made to the two editions of the Panorama of EC Industries that have been published so far; and evidence has been submitted to the Commission in response to a number of their enquiries. We also have an annual meeting with relevant Directorates of the Commission.

So I think we may fairly claim that the Commission is now well aware of the existence of construction economists.

1.2 PURPOSE OF CONFERENCE

What was the purpose of the conference? In part, it was to report what CEEC had been doing and to obtain the delegates' views as to what CEEC ought to be doing. It was also to compare CEEC's experience of pan-European working with that of our colleagues in related professions.

1.2.1 The profession in Europe

A frequent comment in the UK is to the effect that there are no quantity surveyors in the rest of Europe. In fact, there are about as many as there are in the UK. Spain has 16,000 technical architects, and the profession has existed since the sixteenth century. There are 2,000 in Portugal. France has 8,000 construction economists with a professional history stretching back over a hundred years. The Danish Association of Construction Economists, which sprang from the Association of Danish Architects, is now standing on its own feet; special chapters for construction economists exist in Belgium and Holland; in Germany BDB represents construction economists; and in Finland construction economists go under the title of value engineers, though their work is much nearer quantity surveying than value engineering. Construction economists in Switzerland are represented on CEEC.

1.2.2 The work of the profession

Of course, the role varies somewhat according to the local traditions and requirements, but most construction economists offer most of a well-defined range of services including:

- investment appraisal;
- cost control during design;
- estimating;
- preparation of contract documents;
- evaluation of tenders;
- cost control post contract;
- preparation of interim valuations;
- preparation of final accounts.

We can perhaps term these **traditional services**, and they remain of central importance to clients; for our part we must maintain our reputation for solid and reliable work in these fields.

More recently, we have begun to offer a range of services of a value-added character, and these are assuming increasing importance in our workload, particularly of the more advanced firms. These services, which Michael Hartmann, my senior Vice President, calls **soft data**, include:

- life-cycle costing;
- value engineering;
- maintenance economics;
- environmental economics;
- contractual dispute resolution;
- procurement selection;
- project management;
- management of construction programmes.

In fact, construction economists work in just about any field for which their training and education in construction technology, contract law, applied economics and contract administration fits them. The parallel with the spread of accountancy is quite close.

The conference was also a chance for us to share the results of our work programme and, perhaps most importantly, it was an opportunity for construction economists in all European countries to exchange views and experience and, if they felt so inclined, to exchange marriage vows and fix the dowry.

Above all it was three days in Paris in the spring in the company of UNTEC who, as I well remember from their meeting at La Rochelle, are very capable of enjoying themselves.

2

Keynote address

Robert Martin

Ladies and gentlemen, members of the CEEC: I am very pleased and honoured to be here today at this historic First European Conference of the CEEC. While your origins as a European professional organization go back almost 15 years, my own relations with you are much more recent. I was however consulted by your Presidents, both of CEEC and UNTEC, and others during the gestation period of this Conference, and although my own part was minimal, I am none the less extremely pleased to see the concept come to full term here today.

I am also more than a little apprehensive – I see I am billed to give the keynote speech. Not being of the profession of construction economist, nor even an economist of any hue, I wondered how I might fulfil the function you have kindly assigned to me. I decided to seek inspiration from the rest of the programme, thinking that a *bon fonctionnaire* should be able to find something in your proceedings on which he could comment in a knowledgeable way – something to seize on to give you exhortation and intellectual sustenance to carry you through the rigours of the next three days ... and beyond.

I thought that the topic 'The human factor in management' might contain some scope, but we bureaucrats are not famed for our prowess in management: for those versed in these matters I will just say that we are devout followers of the mushroom principle of management. 'European cost data bank'? Given the public perception of the cost of Europe and the budgetary difficulties we seem to find ourselves in perpetually, whether CAP or other, I decided, on grounds of political sensitivity, that this would not be a prudent subject to use as a springboard for my exhortations – the same, by the way, went for 'Exchange of costs', 'Cost evolution', and *especially* 'Global costs'.

'Professional training' seemed to offer scope; there must be something novel I could add to the debate, as an official who has in his time covered about a dozen very different areas of responsibility. But then I realized that, apart from my initial training as a lawyer, I had not received a single day's training during 20 years to adapt to the new responsibilities that I seemed to face every two or three years. I have, none the less, over time, become what the Irish call an 'expert'. With irrefutable logic the Irish consider as an expert any government or other official who travels! As I am an official who has travelled to Paris, I am therefore an expert – QED! I had to conclude however that I hadn't many *bons mots* on professional training to offer.

So, ladies and gentlemen, rather than talk about what I know profoundly little of, I will offer you some thoughts on what I know relatively better – the service sector within the Single Market. To be more precise, I would like to focus my remarks on **the challenge of the Single Market**. I shall deal with two aspects:

1. what the Commission, as a Community Institution, and – more specifically – the new Internal Market and Financial Services DG, is doing to rise to that challenge;
2. what you can do to take maximum advantage of the opportunities offered by that Single Market.

But before embarking on the task of seeing where we wish to go together, it may be useful to set the context by looking at where we are now.

The Single Market became operational on 1 January 1993 and with it the framework for the Commission's activities over the next couple of years. The success of the Single Market will depend on how it is managed and that success, in turn, is itself crucial for the Community's credibility, a credibility that frankly took some knocks in the year that has passed. The Single Market is also the most immediate, practical and visible manifestation of what European integration is about – what it has to offer the businessman, the worker and ordinary citizens. The general public equally expect to avail themselves fully of the four freedoms that are the pillars of the Single Market:

• free movement of goods;

- free movement of persons;
- free movement of capital; and
- free movement of services.

The proper functioning of the Single Market is also critical for economic success. It is vital to make the most of its potential for growth – and herein lies the core of my message for you today.

You will not be unaware of the fact that in recent years European industry, and as a result the European economy, has shown signs of weakness. The indications have been clear: Europe's competitive edge has been blunted, its research potential eroded, and it is not in a strong position with regard to the technologies of the future. The decline in the Community balance of manufactured goods between 1985 (then standing at plus ECU 116 billion) and 1990 (down to plus ECU 50.5 billion) shows how fragile the competitive position of European industry has become compared with that of US and Japanese industry.

Some figures: the Community R&D effort in 1991 was 2.1% GNP, comparable with that of Japan 10 years ago. Now Japan has devoted 3.5% to R&D and the USA 2.8%. Advanced technology goods account for 31% of US exports and 27% of Japanese exports, but only 17% of Community exports.

So much for industry: what of services? Here the picture is mixed. At Community level, according to the latest survey of EUROSTAT of the period of the 1980s, EUR 12 (the balance of trade in services) remained healthily positive throughout the early and mid-1980s, reaching a figure of ECU 13.8 billion (or 5.3% of total trade) in 1989, and this after three years of relative stagnation. The US surplus on services trade for the same year – 1989 – amounted to ECU 23 billion. To complete the picture on a more cheerful note, in Japan where services is not a natural growth sector of the economy in the same way as manufactured goods, the deficit on services of ECU 29 billion in 1988 rose to a record ECU 37 billion in that same year of 1989.

Recourse to these figures is intended to demonstrate what I said earlier concerning the loss by the Community of its competitive edge. They also demonstrate and underline the new priority that is accorded to competitiveness in the Maastricht Treaty, or draft Treaty. This new Treaty makes **competitiveness** a priority for the Community between 1993 and 1997. For the first time, the new Article 130 makes industrial competitiveness, in an open competitive market, a central issue. The provisions of R&D spell out the link between this and other policies, while other measures incite the Community to create infrastructure networks intended to ensure that the Single Market operates effectively.

Competition is now the main driving force behind the changes taking place: maintaining it is the prime condition for success in the necessary process of adjustment. The pressure of competition is urging on a wave of change and restructuring in both the industrial and service sectors: changes in one sector stimulate changes in others. So the task facing the Community – all of us – is to anticipate and implement change on the one hand, and, on the other, to seek to avoid slipping back into national fragmentation and uncoordination. In this context, Community action must complement action undertaken by the Member States and by the business world. We are all part of the same endeavour.

So, in this new competitive climate of the Single Market, what can the institutions do – what are they doing – to adapt to the new challenges? Traditionally the Commission, as the guardian of the Treaty, saw its role as essentially a threefold one:

- it monitored;
- it enforced; and, if necessary,
- it proposed new solutions to new problems,

traditionally by means of Directives. Some of you may recall the old sector-based approach and the difficulties that it engendered when trying to negotiate a proposal through the Council. The Architects' Directive is the classic example: it took 18 years of negotiation, and its example played a significant role in the shift from vertical to horizontal initiatives.

New situations bring about new techniques, new ways of going about old tasks, and new tasks. The Directive as an instrument or vehicle for legislative change has not simply gone from vertical to horizontal, where appropriate, but it can adopt new, more efficient approaches to problems. In the area of standards, for example, the old method consisted of agreeing the content of standards for, say, lawnmower noise levels, painfully and slowly first with Member States' experts, and then in the Council. Now we use the technique of reference to standards that are not themselves negotiated within the legislative process, but are rather entrusted to external bodies such as CEN, CENELEC and ETSI for negotiation and adoption.

Still on the theme of adjustment: **adaptation**. You will be aware of the broad distinction between what I might call the regulated products and the non-regulated products. Within the general context of facilitating the free movement of goods, we attempted to harmonize by means of Directives in those areas where we encountered differing requirements, regulations and measures in the different Member States. But there still existed other obstacles to the free movement of goods, which originated in the **market** rather than in the Statute Book.

I am referring here to procurement practices, and to insurance requirements, which frustrated, or made more onerous, cross-border trade in goods. In this context we developed a new approach for what I call the non-regulated sector, which is contained in the foundation of EOTC (European Organization for Testing and Certification). The purpose of EOTC is essentially to give the European market confidence that tests and certificates on industrial and other products, issued anywhere within Europe, are properly performed so that they do not have to be repeated. We have placed high reliance on EOTC as providing a focal point for conformity assessment in Europe for the future.

Enforcement is another duty of the guardian of the Treaty, which is also encountering the winds of change and adaptation. Traditionally it consisted of recalling to Member States their obligations under the Treaty, and initiating actions before the European Court of Justice where the Commission considered that a Member State had failed to fulfil an obligation. This procedure is less stark in the present climate. Thus the Commission places more emphasis on ways and means to prevent obstacles arising in the area of free movement, whether of goods or of services: for example by recourse to the technique of **mutual recognition**. This is particularly notable, and necessary, in the area of cross-border provision of services.

Despite the provisions of Article 59 of the Treaty, progress on the free movement of services between Member States has been much slower than that relating to goods. The share of services in total intra-Community trade is declining at a time when the macro-economic significance of the sector – particularly in terms of its contribution both to the Community GDP and employment – is increasing steadily in each Member State. To combat this situation, the Court – in an increasing number of cases – and the Commission have resorted to the principle of mutual recognition/home country control, better known in the field of goods than in services. Lawyers among you will know of this principle by its other appellation, *Cassis de Dijon*. This principle states that if a product or, by extension, a service is legitimately put on the market of its home country, there is – at first sight – no good reason why it cannot equally be offered in the market of another Member State.

We in the Commission are engaged in developing further this mutual recognition approach for deployment in the services sector. This is not entirely novel, as the approach has already been implemented in the financial services sector. For example, the Second Banking Directive, which came into force at the beginning of this year, created what is known as a single passport for each bank within the Community. This allows a bank to set up branches and offices in other Member States on the basis of an operating licence issued by its home country. No operational authorization is needed from the host country, although banks must still respect the local rules of basic commercial behaviour. This principle of **home country control** is also at the basis of the key liberalization directives for other financial services and for insurance.

This same approach is good also for great swathes of the service sector, although there may well be some subsectors where Member States could demonstrate that its adoption would not be appropriate.

Another new approach to enforcement, highlighted in the Sutherland Report and now part of the Strategic Plan for the Internal Market, consists of a sharing of the responsibility for the enforcement of Community rules with Member States bodies – a **partnership**. I am tempted to say that this is a case of the poacher obliged to become gamekeeper ... but I will not.

In these changing times, enforcement has evolved with the pressure of new commitments –new tasks in new circumstances – just as I have tried to show that the techniques underlying the legislative process have themselves evolved with time.

Finally, I come to **monitoring**, where my interest is concentrated on the area of service statistics. Statistics represent an indispensable tool for the task of monitoring the economic health of a nation or a Community, and for diagnosing the ailments that strike from time to time. But if the necessary inputs of information and data are not adequate, then the mechanism is of little use; it may even lead to confusion if the diagnosis is inaccurate.

Service statistics are in a deplorable condition, and for a variety of reasons. First, serious interest in the service sector is a very recent phenomenon, which effectively dates back to the inclusion of services in the Uruguay Round in the mid-1980s. It is therefore an area of neglect, which is only now being remedied, albeit gradually.

Second, there is the inherent difficulty of adapting classification and collection procedures developed for, and focused on, finite products to the very ephemeral area of services, where a service does not exist until it is performed, and where it is consumed in that performance: you cannot stock services!

Third, there are the intensely practical problems posed in the recognition and recording of the existence of service activities. By that I mean that if A makes a widget for export to B in another country, the widget arrives at the port and can be observed and counted. But if A communicates some advice to B in that other country by telephone and, just to complicate matters, B reciprocates with advice in some other area some time later, and both parties consider they are 'quits' and no funds change either hands or countries, then the transactions, although real, may well not figure at all in the export/import statistics – such invisibles they truly are!

To bring some order into this chaotic situation, Eurostat – the Statistical Office of the European Communities – has been working for some years now to create a proper framework for service sector statistics.

This resulted in a Council Decision, adopted in June 1992, which established a two-year programme for the development of Community statistics on services. One of the many aspects of this Herculean endeavour relates to the drawing-up of a methodological manual, which among other things, tackles such vexed areas as the definition of, for example, variables, statistical units, and orders of magnitude, as well as the area of nomenclature. To record or register one must be able to identify an activity, and thus the question becomes one of defining clearly and exhaustively what is done by a construction economist, so as to register those activities *separately* from those carried out by, for example, architects: how may one distinguish the activities of legal adviser, accountant, management consultant each from the other?

We have also established a European statistical information system, or data bank. This is called MERCUREY: very appropriate when you consider that he was the Roman god of industry and trade – somewhat less so when you realize he was the protector, not only of public speakers, but also of gamblers ... and thieves.

One final remark in this context: the task of creating from scratch a single new statistical corpus is a daunting one, but when that task involves – as it must do in a Community of 12 Member States – the bringing together and harmonizing of the statistical practices of 12 national statistical offices unused to cooperation with anyone, it assumes gargantuan proportions.

Allied to this activity of building up a general horizontal diagnostic and monitoring tool, my Unit within the Commission has been undertaking a long series of sectoral studies aimed at presenting an economic analysis of activities largely contained in what is usually called the business services sector. The latest of these, which will be published shortly, is entitled *The Legal and Economic Aspects of Management Consultancy within the Community*.

Another type of monitoring activity of which you will be aware is the publication, by another service in the commission, of PANORAMA, the Community-wide survey of industrial and service sectors. The 1993 edition will appear shortly, around the beginning of June, and will feature a monograph on construction economists: alas, I must add, still yoked to architects. The good news, however, which I am pleased to announce to you, is that you will *again* have your own monograph in the 1994 edition, which we have already begun to work on.

Enough then of the Commission, and on to some thoughts as to what *you* might do to rise to the challenge of the Single Market: how you might extract maximum advantage of the opportunities available.

The first, and by far and away the most obvious, step is to realize that you are now operating within the Single Market and no longer in one, or several, separate, national, markets. Clearly, the proof that you have the right reflex for this new attitude of mind resides in the fact that we are gathered here today at your first *European* Conference, in this the first year of existence of the completed Single Market. But to operate successfully in as large a playing field as the geographical territory of the European Community requires a transformation of mind set: it is no longer sufficient to consider the *national* market as the theatre of operations. The sheer magnitude of this new Single Market should inspire you – compel you – to compete in each and every part, so that the end-result is that the most efficient practitioners will succeed and, in turn, act as a spur to the less successful, to the overall benefit of all. The economic success of both the USA and Japan is derived *directly* from the availability of a large integrated home market. If we, in Europe, are to achieve our ambitions of improving our standards of living, the very first step has to consist in coming to terms with the existence of a similar internal market; the second is to set about its exploitation in a vigorous and competitive spirit.

Competition does not, however, exclude cooperation. The cooperation that you have developed over your 15 years of existence has enabled you to develop your profession within the European dimension in such areas as training and qualifications, practice, and coordination of working methods, as set out in the CEEC objectives. A stronger, more self-reliant, profession can only result from such activities, which are founded upon the notion of cooperation between your constituent parts.

Cooperation at European level, virtuous though it is, should not however be taken to excess. By that I mean that excessive attention to self-betterment and self-promotion *can* lead, in certain circumstances, to a form of protectionism. The temptation is ever-present to seek to secure and defend **exclusive rights**, whether to certain activities or to designated territories, by the aggressive development of what the French call *déontologie*: the ethics, the practice guidelines that any reputable professional organization draws up in its entirely laudable efforts to raise standards of performance and behaviour.

I would urge you, as I do all professional organizations, to be vigilant lest excessive zeal in the area of *déontologie* leads to the *de facto* erection of new barriers to trade, just at the time when all our efforts are deployed to dismantle those that remain in the way of a truly Single Market.

Reference to *déontologie* and ethics leads on naturally to the subject of **quality**. In the business world of today, in the economic climate of today, you must compete or decline: the key to success in competitiveness undoubtedly lies in quality.

I spoke earlier of our new emphasis, in the Commission, on quality, and the reliance we have placed in EOTC in the non-regulated sector. I would stress that EOTC is not simply a mechanism for use in the goods sector, but it is equally applicable, equally capable of exploitation, in the services sector. Indeed, the very first EOTC Recognized Agreement Group related to an accreditation system for calibration laboratories – very much a service activity. We would be only too pleased to assist you, with advice rather than funding I would add, in any ideas you may have for programmes dedicated to the improvement of standards of quality.

And so finally I come, almost full circle, to the subject of statistics: one that I note figures among the aims of CEEC. As I tried to demonstrate earlier, statistics play a vital part in the monitoring, and diagnostic, activities of the Commission. But while our role as public sector operators is to create the framework and devise the methodology, such is the nature of service activities –largely in the hands of the private sector – that we must rely on you to fill in the blanks, the empty spaces, and so make sense of what, in the absence of your cooperation, would remain just that: a framework, devoid of content.

Some of your officials attended the one-day services sector meeting that I organized last December in Brussels and, perhaps, were taken aback by the fact that a full half of the proceedings was devoted to statistics. We said then, and I make no excuse for repeating it here today, that the Commission's statistical programmes need, and cannot be implemented without, the active cooperation of sector bodies such as yourselves in the collection of the raw data. It is, I know, a sensitive area, and one always fears that data put into the public sector for one purpose may end up being exploited for another. I can only say that the greatest degree of confidentiality is accorded to such submissions: if this were not the case, we could never begin to function. I must therefore ask you, against this background, to give us the benefit of your best endeavours in the area of statistics.

And so, in conclusion, I would say this: our activities in the Commission over the next couple of years will, as I have said, focus on:

- stimulating the economy;
- strengthening economic and social cohesion; and
- boosting business competitiveness.

In all frankness, this last objective is the *sine qua non* of others. Unless we succeed in the double objective of halting our economic decline and, at the same time, making a positive success of exploiting the Single Market through a renewed dedication to competitiveness, we will not have at our disposal the wherewithal to achieve our other, broader, objectives.

Only with the full support of the business community, of groups of dedicated professionals like yourselves, can we make it to the starting line. We have to renew our faith in partnership: between the public and the private sectors.

Thank you, ladies and gentlemen.

3

Developing the role of the building surveyor within the European Community

John Pelling

The European Community is of course about working together with our European neighbours to create a market for business that will match those of the Atlantic and the Pacific. It creates threats and opportunities to both the well-known and lesser-known European professionals, and you have already heard from the largest group, the architects and engineers. This chapter is concerned with one of the lesser-known groups. It is intended to try and explain the role of the **building surveyor** to our non-UK colleagues, and to set out the objectives of our Pan European Group. In outlining the development of the Pan European Group for Building Surveying I need to set the background.

The term 'building surveyor' will probably not be known to many of you, and, indeed, as a description of the function of the people who call themselves building surveyors it is misleading if literally translated from the English into any of the other European languages. The building surveyor is somebody educated and trained to be an expert in the technological and management processes by which buildings are repaired, refurbished and constructed. Traditionally his principal activities have been directed at existing buildings rather than new buildings.

In the UK, building surveyors obtain their specific technical and management education and training alongside a background knowledge of land economics and property and contract law. They will also be educated alongside other construction and property industry professionals. There are approximately 12,000 building surveyors in the UK, either qualified or in the course of training. Building surveyors spend three years at university and two years in practical training within the professional working environment, and at the end of that two years take an examination by their peers. There is therefore a minimum period of five years of education and training.

When we started our research I tried a comparison between the UK and the remainder of the European Community. Figure 3.1 shows the number of typical first-degree courses in the area of the built environment available in the UK simply compared with those in the rest of the European Community.

However, within the European Community context, who else has the pool of knowledge available to the UK building surveyor and practises in similar areas? We shall not find the answer to this question by looking for the title 'building surveyor' in the wider European context. It is necessary to look at function. Our initial research indicated that a similar function to that of the building surveyor was carried out by individuals who had been educated first at university or college as architects or engineers and, in some countries, as geometers. They had later specialized by practice and with further study.

As Fig. 3.1 indicates, the UK approach differs from this latter process of a first degree in architecture or engineering followed by a specialism in practice. The UK has a number of first-degree courses for particular specialisms of the built environment. Hence specialism starts at a far earlier stage. There are obviously benefits in the wider European system, but I suspect that it does not create the same opportunities as the specialist UK courses, and tends to suppress competition amongst the existing and emerg-

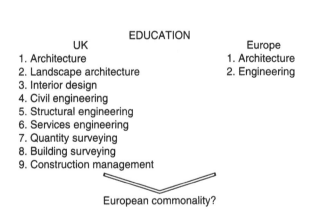

Fig. 3.1 Typical first-degree courses in the UK and the rest of the EC.

Table 3.1 Schedule of key comparisons 1986/87

Country	GDP (million dollars)	Construction (% of GDP)	Population per km²	Population (pop. in constr.)	Percentage of population in construction	Percentage of national construction output Civil engineering/ building
France	489,000	11.0%	103	57.0M (1.6M)	2.8%	23/77%
West Germany	613,000	13.8%	249	61.0M (1.4M)	2.3%	20/80%
UK	425,000	8.5%	231	56.5M (1.5M)	2.6%	20/80%
USA	3,500,000	9.0%	25	239.3M (6.1M)	2.6%	25.5/74.5%

Source: DOE Report *Professional Education for Construction 1989*

ing professions. The number of specialist UK courses also provides practical education difficulties, and I would see some commonality arising.

A study was carried out in 1989 on behalf of the UK Government's Department of the Environment, which compared professional education for construction in the UK, France, Germany and the USA. Table 3.1 shows the position of construction in the overall economy, and for this purpose I am comparing only France, Germany and the UK. You will see that as a percentage of gross domestic product there are some differences, but the actual percentage working in construction is very similar, varying from 2.3% to 2.8%.

Table 3.2 shows the split of the professionals working in each of the countries. It lists the numbers of architects working in each country and the numbers of building engineers in France and West Germany. Under the heading 'building engineers' in the UK it substitutes the various professionals, and building surveyors are listed there alongside the other construction professionals. The figure of 12,000 building surveyors given earlier is more up to date, and includes not just those who are fully quali-

Table 3.2

Country	Registered architects	Architects per 1M inhabitants	Number of building engineers (including surveyors and builders)		Building engineers per 1M inhabitants
France	23,000	404	Building engineers	23,000	404
West Germany	60,424	990	Building engineers	100,000	1639
UK	30,475	531	Quantity surveyors (RICS)	21,055	372
			Building surveyors (RICS)	4,218	75
			Builders (CIoB)	9,542	169
			Services engineers (CIBSE)	7,012	126
			Structural engineers (ISE)	11,179	198
			Civil engineers (ICE)	49,457	875
				102,436	1815
USA	60,000	250	Building engineers	341,000	1427

Source: DOE Report *Professional Education for Construction 1989*

fied, but those who are currently working towards becoming qualified. It also includes non-RICS building surveyors.

The building surveying profession has grown substantially in size and reputation in the UK in the last 20 years. This has not just happened by chance. The well-documented change in activity from redevelopment of our cities and their buildings to repair, refurbishment and construction, which started in the late 1960s and early 1970s, required the traditional skills of the building surveyor as a pathologist and technologist of existing buildings: skills that were necessary to start and sustain repair, maintenance and refurbishment programmes.

It is against this background that an initial meeting of individuals from Belgium, Ireland, the Netherlands and the UK, representing individuals who carried out a similar function, was held in 1990 to discuss the formation of a Pan European Group. Common areas of activity were discussed, and it was decided that there should be a body to represent this branch of property and construction professionals within the European Community. Encouraged by your CEEC President Brian Drake, and by Department DGIII of the European Commission, discussions took place in 1990 and early 1991 and culminated in the formation of the Association d'Experts Européens du Bâtiment et de la Construction (Association of European Building Surveyors). I was privileged to be elected its first President at that time.

Six countries are now represented within the AEEBC. Each country can be represented by a number of organizations, but can only have three delegates at the twice-yearly General Assemblies. Belgium is represented by the Union Belge des Géometres Experts Immobiliers. France is represented by three organizations: the Compagnie Française des Experts Construction, the Ordre des Géometres Experts Français, and the Union Nationale des Techniciens de l'Economie de la Construction. Those three organizations I believe have now come together into a new umbrella group to represent their interest. The Netherlands is represented by the Nederlandse Vereniging van Bowkostendeskundigen, Ireland by the Society of Chartered Surveyors in the Republic of Ireland, and Italy by the Consiglio Nazionale dei Geometri. The UK is represented by The Royal Institution of Chartered Surveyors, the Association of Building Engineers, and the Architects and Surveyors Institute.

There are of course some omissions, and in particular we would welcome membership of our organization from the technical architects in Spain and Portugal and our Danish construction economist colleagues. Germany has proved a difficult area in which to identify a group for membership, but is obviously a very high priority on our list for membership.

The common activities of this European grouping of property and construction professionals are the technological and management processes by which buildings are maintained, repaired, renewed and constructed. The specific areas of activity are based on the overall activities of our members, but it does not follow that all carry out all activities in all of the European countries. The specific areas are:

- building pathology;
- building asset protection;
- building maintenance and repair (design, supervision and specification);
- building refurbishment and rehabilitation (design, supervision and specification);
- building control;
- building project management;
- building experts.

Individuals represented by the association work across all sectors: central and local government, utilities, retail, industrial, commercial, construction, development, and of course as consultancy practices. The consultancies provide service to all of the other sectors.

I should like to go through a few slides, which I think will indicate more clearly some of the building types that we work on, and the activities of our members.

This is social housing in an inner area of London belonging to a local authority, and demonstrates one of the types of building on which we carry out our **building pathology** activities. Here we have a concrete-framed and concrete-panelled building in which we investigated the defects in the concrete panel system and made recommendations for resolution of defects, designed and specified the solutions, and finally monitored the subsequent remedial construction. The next slide is a similar project, again a building built a little later in the 1970s. It has a similar concrete frame and concrete panels, and the same services as on the last project, but this building belongs to a very large UK property investment company.

This is a rather splendid late-Victorian house, in one of London's 'leafy suburbs', and demonstrates a project on which we would carry out the services of **building asset protection**. In this case the house was proposed to be purchased by a private individual and the building surveyor carried out a condition survey, prior to purchase, to advise the proposed purchaser on his liabilities. Here is asset protection on a different scale: a project constructed in the South West of England by a developer, but with an insurance company investor. Here the building surveyor monitored the activities of the developer and his design team on behalf of the investor. He would be looking at both building quality management and also building risk management.

Building maintenance and repair is a fundamental activity of the building surveying profession. It includes maintenance investigation, maintenance planning and maintenance management. Information technology is used both to collect data and subsequently to analyse and prepare recommendations. This is a slide from a magazine article referring to maintenance investigation and planning carried out on behalf of local education authorities. This is a similar exercise on operational buildings owned by London Transport.

Building refurbishment and rehabilitation is again a mainstream activity of the building surveying profession and in certain of our European countries the building surveyor will be involved in the design, specification and management of the schemes for the alteration, refurbishment and rehabilitation of existing buildings. This is a slide showing CAD systems at work. The following is a rather fine Grade I listed building converted into residential units and on which the building surveyor carried out the full range of activities: as the designer, specifier and supervisor of the refurbishment works. This is another, contrasting, building in which the building surveyor was the designer, specifier and supervisor of all of the refurbishment works to this concrete-framed and concrete-panelled social housing block of the 1960s. This is a similar scheme: this time internal office refurbishment on behalf of a property developer.

Most of the slides that I have shown you so far involve some sort of construction activity that requires building control. Building control in the UK is a function currently carried out by local government officers, although there is legislation in place to enable control also to be provided by the private sector consultant. At the moment that legislation has not been carried through because of the lack of insurance to cover those people who want to become private-sector building controllers. Therefore most building control is in the hands of the local authorities, except for one or two very large organizations such as the National House-Building Council. Building control officers have again traditionally come from the ranks of building surveyors. In addition, control is often carried out on behalf of an investor, and the next slide shows a factory unit in which the design and construction phase was monitored and controlled by building surveyors on behalf of an investor.

Increasingly in the last ten years building surveyors have become involved with new buildings as project managers, managing the activities of the design team on behalf of clients. This slide shows part of a new industrial development and the next slide an office refurbishment carried out on behalf of the UK Government's Property Services Agency.

As we all know, some building projects and some building design teams tend to create problems, which results in litigation between the various parties. It is in this area that the building surveyor can operate as a building expert and seek to act on behalf of the various parties.

Finally, I should like just to apologize to some of my colleagues that all the slides are of buildings in the UK, but I assure you that this is merely a case of practicality for this particular seminar.

These then are the principal activities of the members of the AEEBC. Not all of the members will carry out these activities: for example, they will not be allowed in some of the member countries to carry out any of the design function on refurbishment or new-build projects. However, in those same countries they may also have a much higher profile as building experts. We have therefore a broad base of expertise, but also highly skilled expertise in particular areas. We believe that the basic education and training of building surveyors in all European countries, leading to their expert knowledge of construction technology, gives room for development of this profession along the lines of its development in the UK during the last 20 years.

That conveniently leads me on to the aims of the Association, which are:

- to represent the interests of its members to the European Commission and other European institutions;
- to facilitate education, training and mutual recognition;
- to establish professional codes of conduct;
- to establish practice guidelines.

Education, training and mutual recognition are of fundamental importance to our Association. In the short term we aim to develop postgraduate courses in our specialization to add to existing European courses in architecture, engineering, geometry, and construction economics. We are also conscious of the need for continuing training and development of our qualified professionals during the course of their careers to give them the skills needed to work in an industry of rapid development.

As a European group we have a specialist and recognizable skill. It is a skill that we believe has become important over the last two decades and whose importance will continue to grow as the need grows for care and protection of our existing building stock.

4

The impact of the Single Market on construction

Michael Webb

4.1 INTRODUCTION

The potential impact of the Single Market on construction is very broad, and almost impossible to forecast with any degree of accuracy. However, I come to Paris as a humble Irishman in the footsteps of more distinguished predecessors like James Joyce, Oscar Wilde and Samuel Beckett, who came to France to escape the (then) insularity of Irish society. Thanks to our membership of the EC, and the fact that half our population is under 25 years of age, that insularity has been softened as Ireland grows and develops as a society and as an economy.

Samuel Beckett was the son of an eminent Irish quantity surveyor/construction economist, whose practice still continues today. It was in the attic of his father's quantity surveying office that the young Samuel Beckett first began to write – writing that blossomed when he moved to live in Paris.

But, as Vladimir says in *En Attendant Godot*, 'Come, let's get to work'.

4.2 CURRENT SITUATION

I should like, first of all, to examine the current situation with regard to cross-border trade within the EC, and then move to examine how the Single Market will affect that trade.

4.2.1 Construction materials

Over the past decade the manufacture of construction materials within the EC has grown from 36 billion ECUs in 1980 to 59 billion ECUs in 1990 – an increase of 65%. Over the same period in the USA the manufacture of construction materials increased by 75% to 36 billion ECUs in 1990.

It is in trading those materials that the most dramatic changes have taken place. Exports from the EC increased by 60% over the decade to 2.8 billion ECU, while imports to the EC increased by a dramatic 220% to total 1 billion ECUs in 1990. However, it is within the EC itself that the most interesting changes have taken place.

In 1980, intra-EC trade totalled 1.8 billion ECUs. By 1990, thanks to the gradual relaxing of barriers to trade, intra-EC trade has increased to 4.4 billion ECUs – an increase of 140% in 10 years (Table 4.1). In 1980, cross-border trade accounted for 5% of the total production of building materials. Ten years later, cross-border trade had risen to 7.5% of the total. Over the same period imports from outside the EC rose from 0.8% (1980) to 1.6% (1990).

For the most significant building material – cement – the trends were similar. Cross-border trade within the EC accounted for 3% of apparent consumption in 1980 and rose to 4.5% in 1990. However, over the same

Table 4.1 Intra-EC trade in construction materials

	Total production (million ECU)	*Intra-EC trade (million ECU)*
1980	35,792	1,819
1981	37,614	1,769
1982	38,444	1,899
1983	40,829	2,075
1984	42,111	2,302
1985	41,459	2,406
1986	41,657	2,673
1987	44,256	2,931
1988	50,677	3,484
1989	55,725	4,029
1990	59,077	4,387

Source: Panorama of EC Industry 1992 Supplement

period imports from outside the EC rose by 650% to 6 million tonnes (or 3.5% of total production), while exports from the EC to outside countries fell by 130% to 9 million tonnes (5% of total production).

With the further opening of markets and removal of barriers, how much more dramatically will intra-EC trade grow in this decade?

4.2.2 Contractors

The total output of the EC construction industry was 460 billion ECUs in 1989, which accounts for more than 10% of the EC gross domestic product.

The relative size of the construction industry in each of the EC countries is important in understanding the potential (Table 4.2). However, when one looks at the top 30 EC contractors by turnover, these differences are accentuated. Obviously some of these companies are conglomerates and construction is but part of their empire – that said, the comparisons are of interest (Table 4.3). Of the top ten companies, six are French, three are

Table 4.2 Construction output – 1991

	Billion ECUs	*%*
Belgium	17.8	4
Denmark	12.5	3
France	79.2	17
Germany	125.9	27
Greece	5.8	1.2
Ireland	3.4	0.7
Italy	76.1	16
Luxembourg	1.0	0.2
Netherlands	23.1	5
Portugal	6.8	1.2
Spain	42.3	9
UK	68.7	15
Total	461.6	100

Source: EC Panorama 1991

British and one is Belgian. Of the top 30 companies, ten are French, nine are British, while only four are German, three Spanish, and one each is Belgian, Dutch, Danish and Italian. Each of the top ten companies has a turnover in excess of the total output of the whole Irish construction industry.

Table 4.3 EC contractors 1992

		Country	*Million ECUs turnover*
1.	Generale des eaux	France	16,049
2.	Lyonnaise des eaux - Daumez	France	12,543
3.	Bouygues	France	8,972
4.	Schneider	France	7,215
5.	SGE	France	5,620
6.	Tractebel	Belgium	5,252
7.	BICC	UK	4,926
8.	Tarmac	UK	4,602
9.	Trafalgar House	UK	4,574
10.	SAE	France	3,908
11.	Philipp Holzmann	Germany	3,857
12.	GTM	France	3,707
13.	AMEC	UK	3,336
14.	Spie-Batignolles	France	3,330
15.	Formentro de Construcciones	Spain	2,836
16.	Dragados Y Construcciones	Spain	2,834
17.	Wimpey	UK	2,411
18.	Laing	UK	2,251
19.	Hollandesche Beton	Netherlands	2,057
20.	Fougerolle	France	1,995
21.	Taylor Woodrow	UK	1,990
22.	Mowlem	UK	1,978
23.	Costain	UK	1,877
24.	Cubiertas Y Mzov	Spain	1,790

Table 4.3 (continued)

		Country	*Million ECUs turnover*
25.	Bilfinger & Bergen	Germany	1,734
26.	Strabag BAV	Germany	1,650
27.	Colas	France	1,625
28.	Dyckerhoff & Widmann	Germany	1,598
29.	FSL	Denmark	1,593
30.	Autostrade	Italy	1,424

Source: *International Management* – February 1993

It is also relevant to note that 91% of EC construction firms employ less than 10 people and only the top 0.5% employ more than 500 people. Most construction firms are therefore locally based and serve a limited geographical area. The medium-sized firms may operate on a regional basis but it is only the really large firms that operate on a national basis. Likewise, it is the really large firms that are the only ones likely to trade across national borders in a significant way.

The EC has no statistical data on the volume of cross-border trade by contractors, so we can only assume that it is not a significant factor within the EC economy. We do, however, know that of the 15% of EC GNP that is taken up by public procurement, 2% is outside national borders.

4.2.3 Professions

Building professionals form a much smaller and more disparate range of enterprises within the EC. There are 234,000 architects practising within the EC and 175,000 architectural students (Table 4.4). In Greece there are 1,224 architects per million inhabitants while in Netherlands there are only 313 per million. However, when compared to output, the Netherlands has the fewest or most efficient architects with only 202 per 1 billion ECUs of construction output, while Greece heads the league table with 2,110 (Table 4.5).

With consulting engineers the situation is similar. There are a total of 171,000 consulting engineers working in construction (Table 4.4). Their relative efficiency can be seen from the range of 222 engineers per 1 billion ECUs of construction output in Germany to 725 in Denmark (Table 4.5).

Table 4.4 EC construction professionals

	Architects	*Engineers*	*Construction economists*
Belgium	8,761	5,150	NA
Denmark	5,700	9,060	50
Germany	67,533	28,000	NA
Greece	12,240	3,200	NA
Spain	19,243	14,820	15,000
France	25,746	27,880	6,100
Ireland	1,300	630	810
Italy	53,300	21,500	NA
Luxembourg	265	384	NA
Netherlands	4,665	9,400	5,400
Portugal	4,198	1,960	2,000
UK	31,000	49,260	19,000
EC	233,951	171,244	48,360

Source: EC Panorama 1992

The spread of construction economists is unfortunately fewer in number – doubtless they make up for lack of numbers by their quality! According to CEEC statistics there are 48,000 construction economists in the EC, with the biggest number (19,000) in the UK (Table 4.4). However, as we have seen from Table 4.5, their number compared to total construction output ranged from 4 per 1 billion ECUs in Denmark to 355 in Spain, although in Spain many of these would be 'technical architects' practising in other areas.

Statistics with regard to cross-border practice are not available but would probably show a very small proportion of intra-EC trade.

Table 4.5 Number of professionals – per billion ECUs construction output

	Architects	Engineers	Construction economists
Belgium	492	289	
Denmark	456	725	4
Germany	536	222	
Greece	2,110	522	
Spain	455	350	355
France	325	352	77
Ireland	382	185	238
Italy	700	282	
Luxembourg	265	384	
Netherlands	202	407	234
Portugal	724	338	345
UK	451	717	276

Source: EC Panorama 1992

4.3 SINGLE MARKET MEASURES

Having looked at the present position with regard to cross-border trade for building materials, contractors and building professionals, it is time to consider how the Single Market may affect the present situation.

The range of Single Market measures, now part of EC law, attempts to remove some of the legal and administrative barriers to cross-border trade. The Construction Products Regulations, the CE Mark, the development of EC technical specifications, and of course the removal of customs barriers, all attempt to assist EC manufacturers to sell their products in different markets. The Directives on Public Procurement and the developing situation with regard to construction liability will help to open public-sector markets across the EC. The Directive with regard to the Mutual Recognition of Diplomas, the Public Procurement Directives and the proposed Directive with regard to intra-national partnerships will each help those construction professionals who wish to sell their services outside their own country. Each of these Directives will help to ease the situation. However, other barriers remain.

Construction is of its nature not easily transportable and is very much site-based, although the trend has been towards more and more off-site fabrication. Language remains a major barrier to the export of services and contractors. Another major barrier is the selfishness with which individual countries defend their own self-interest. For example, the EC decisions with regard to the ERM have been sabotaged by the Francfort policy and the determination of the Bundesbank to pursue its own self-interest while creating chaos in the monetary system at large and disastrous consequences for smaller countries such as Denmark and Ireland. But perhaps the major obstacle is that of petty bureaucracy and inflexible administrative procedures, together with our natural conservatism in sticking with products, contractors and professional advisers that we know and have used before.

Our Belgian colleagues in CEEC have prepared a report on such barriers as they affect the Construction Economist [Report on real (as opposed to theoretical) obstacles to the practice of construction economics within the EC].

4.4 TRADE POST-1992

Where does the future lie?

A survey carried out by CEEC among its members shows a consistent view of the effects of the Single Market (Table 4.6). In building materials, where most intra-EC trade already exists, CEEC respondents envisage marginal change. For professional services they expect little or no change, while for contracting they expect marginal change.

Based on these findings and the data available with regard to the present situation one can only speculate as to the future. For building materials the trend over the past decade has been for more and more concentration of manufacture, with a concurrent expansion in cross-border trade. As we have seen, intra-EC trade in construction materials has risen from 5% of the total in 1980 to 7.5% in 1990. It is likely that this trend will continue, and, encouraged by the Single Market measures, at a quicker pace. By the year 2000 up to 15% of all building materials may be purchased from other EC states. It is also likely that the major EC manufacturers will increase their share of EC markets through acquisition of local manufacturers and distributors.

Table 4.6 CEEC – The EC Single Market and construction response to preliminary questionnaire. (Replies received by 14 August 1992.)

Country	Ireland	UK	Belgium	Germany	Denmark	Netherlands	Spain	Italy	Finland	France	Switzerland	Portugal
1. Building materials												
Now imported												
0–25%				*			*	*	*			
25–50%	**		**		**	**						
50–75%												
75–100%		****										
Change expected												
No change									*			
Marginal	**	**	**	**	**	**	**	**				
Significant												
2. Building components												
Now imported												
0–25%			*	*		*	*	*	*			
25–50%	**	**			**							
50–75%												
75–100%												
Change expected												
No change									*			
Marginal	**		**	**	**	**	**	**				
Significant		****										
3. Professional services												
Now imported												
0–25%	*	*	*	*	*	*	*	*	*			
25–50%												
50–75%												
75–100%												
Change expected												
No change					*				*			
Marginal	**	**			**		**	**	**			
Significant			****									
4. Contractors												
Now imported												
0–25%	*	*	*	*	*	*	*	*	*			
25–50%												
50–75%												
75–100%												
Change expected												
No change									*			
Marginal	**	**		**	**	**	**	**				
Significant			****									

For building contractors the existing situation with regard to intra-EC trade is much weaker and apart from major projects is likely to remain very much the same. One might speculate that the existing major players in the EC may very well decide to spread their influence throughout the EC by way of acquisition rather than organic expansion.

With regard to professional services the situation is also developing slowly. There are few physical barriers to exporting services, although the technical ones should not be underestimated. I would speculate that cross-border trade in professional services offers the most potential for growth. Some of this growth may be through cooperation and joint ventures between professionals from different countries. That potential will only be realized if the EC and bodies such as CEEC play their full role in lowering the technical barriers through

- recognition of diplomas;
- an EC qualification;
- more transfer of technical information;
- more opportunities for members to meet and make contacts;
- greater opportunities for students and graduates to learn something of the work in other countries.

The future holds great possibilities if we have the courage to grasp them.

To conclude, I return to my compatriot, the son of a quantity surveyor, Samuel Beckett, and *Waiting for Godot*. Vladimir says, 'That passed the time'; to which Estragon replies, 'It would have passed in any case'. Vladimir has the last word: 'Yes, but not so rapidly'.

5

The construction economist in Europe: roles and remuneration

Martin Wade

The European Committee of Construction Economists (CEEC) was formed some 14 years ago, recognizing that the function of advice on construction costs is a European-wide activity. France, Ireland and the United Kingdom were founder members of the Committee, demonstrating the advanced state of the profession in these countries, and it is significant that nine members of the EC States now have representative delegations on CEEC.

The *raison d'être* for CEEC is twofold: to act as the catalyst for a united voice representing construction economists in Brussels; and to act as a Forum for exchanging information on national techniques and procedures with a view to harmonizing these wherever practical and possible. It may surprise you, or not as the case may be, that this latter activity has proved very difficult actually to achieve. Europe comprises a number of separate nations, each with its own particular and deep-seated history and culture. Until very recently they acted as entirely independent nations, many with closer ties with non-European nations, through colonialism, than with their immediate cross-border neighbours.

While there is a great similarity of purpose between construction economists throughout Europe, there is also a great disparity of procedure and liability. Our historic and cultural backgrounds have an immediate influence; English law, for example is based on Common Law, legal precedent and statute, whereas the French legal system is based on Roman Law and set out in statements of general principle in five Napoleonic Codes. Terms and translations cause difficulties, as evidenced by the phrase 'cost control', which in English means 'controlling costs' but when translated into

French, *contrôle des couts*, means 'monitoring costs'. In the *Starbou* in the Netherlands (the architects' publication setting out the various stages of a project), 'complete design' refers to conceptual design and therefore is subject to the development of working drawings, whereas in the UK the RIBA Stage E 'complete design' refers to a fully detailed design capable of being built. 'Main contractor' in Germany is often understood to mean what in the UK is referred to as a 'turnkey contractor'. And so the misunderstandings in translation continue. The methods of letting building contracts vary from country to country. For example, a coordinating subcontractor in the Netherlands or Germany undertakes many of the roles of the main contractor in the UK, but without apparently accepting many of the responsibilities and liabilities.

It will be seen therefore that a direct comparison of construction industries of the various member nations is fraught with dangers and pitfalls. You think you understand what the other chap is saying – you understand the translation – but the reality is that the local meaning is different. There is urgent work to be done in compiling a glossary of European terms and definitions that are correctly translated in both language and meaning. Indeed, this is essential if cross-border tendering for subcontract packages on an individual construction project can become a reality.

It is important that this very serious issue is stressed and understood before embarking on a review of the work of construction economists and their remuneration within the Community, because while there are many obvious similarities and differences in their activities there are also many unidentified disparities in activity and liability, which make a direct comparison almost impossible. For this reason, and conscious that the majority of the original audience for this chapter was probably French, I have restricted my chapter to describing the activities of the British chartered quantity surveyor, and the likely fee that these activities would attract.

In order to undertake this review, it is sensible to consider what is meant by the title, the education required to become a chartered quantity surveyor, the duties performed, their responsibilities and liabilities, their position within the professional team, to whom they are accountable and last, but no means least, their fee.

It should be noted that while I am looking at this as the chartered quantity surveyor as a consultant, a good number work for contractors, subcontractors, developers, local and central government, public utilities and of course academia.

The British chartered quantity surveyor is in fact a specialist discipline within a more broadly property-based professional body, The Royal Institution of Chartered Surveyors (RICS). Quantity surveyors represent about 40% of the total membership, with the other largest discipline being valuation and building surveyors, loosely banded together as general practice surveyors.

The RICS used to set its own examinations, but members now entering the profession will do so by means of a university degree or college diploma of a standard comparable to that of an architect or engineer. They are required to gain some three years' practical experience in a professional or commercial office and then submit themselves to an assessment of professional competence, which they must pass before they can call themselves a chartered quantity surveyor. During their professional life the RICS insists that they update their professional knowledge by undertaking 60 hours of 'approved learning' over a three-year period, known as continuing professional development (CPD).

The chartered quantity surveyor is therefore a professional of similar educational standards as an architect or civil engineer. Anybody can call themselves a quantity surveyor and practise as such, but it is only members of the RICS who can call themselves 'chartered'; and so it is the chartered quantity surveyor (CQS) that I shall consider in this chapter. Because of the specialist training in construction techniques, together with legal, valuation, quantification and negotiation skills, the CQS is able not only to operate as the construction economist but also as a project manager or commercial developer etc. It is however the role of construction economist that we shall consider here.

The CQS is usually appointed by the building promoter and will have his or her fees paid direct: that is, not through a third party such as the architect or project manager. Standard fee scales are now becoming obsolete, as the last edition published by the RICS was in 1988. It is unlikely that the UK Government would allow another edition to be published, and it is probable that the 1988 edition would be banned if updated. Fees are now either negotiated, as between principal and CQS, or won in competition. For this reason remuneration becomes a fairly difficult matter to compare with other EC construction economists, as it is often dictated by market conditions. Indeed, one of the tasks abandoned by the CEEC was a comparative study of fees across Europe. It failed through the difficulty of comparing our respective tasks, and because in a number of countries 'fee scales' no longer apply.

What is probably more apposite is a comparison of man-hours to resource various activities, identifying associated liabilities and risks, and that is how I wish to illustrate the CQS's role and remuneration in this chapter. To do this I shall assume a new speculative office building of some 1500 m² constructed in a major city-centre location to high standards. The CQS will perform traditional activities as cost manager, with no other responsibilities such as project management. I shall not include items such as taxation advice, life-cycle costing, or the resolution of any contractual disputes that might occur. These are all CQS activities that most firms are able to address outside their main commission.

1000 m² high quality office block
Major city location

Design team comprises project manager, architect, quantity surveyor, structural engineer, services engineer (Fig. 5.1).

Assumptions:

Budget cost:	£10 million (ECU 12.5m)
Pre-contract period:	9 months
Contract period:	18 months
Procurement strategy:	Prime contractor

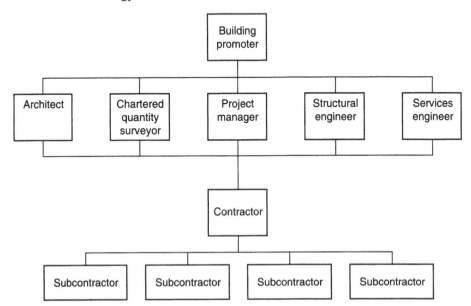

Fig. 5.1 Traditional British design team.

Activity	Man-hours	Comment
Pre-contract		
Receive brief from building promoter	15	
Create cost model based on conceptual design information	100	CQS will take a high degree of responsibility/ liability
Develop cost model into cost plan as design develops including undertaking cost studies of design options	250	Ditto albeit dependent on input of design team
Advise project manager on procurement strategy	25	CQS is usually very involved with this activity and so takes high degree of responsibility
Prepare cash flow forecast and regularly update	100	Ditto
Prepare tax planning exercises	Not included	This is often required, particularly by commercial clients
Prepare 'costs in use' exercises	Not included	This is becoming more often required, particularly by public sector clients
Attend design team meetings and contribute to design development	200	CQS is actively involved during design process, contribution from value for money and cost viewpoints
Liaise with tenant's advisers for pre-let tenant's requirements	Not included	CQS is usually required to enter into a tenant's duty of care agreement (as well as a fund's duty of care if appropriate)
Prepare fully measured bills of quantities	1700	An extremely labour-intensive activity, breaking the design into detailed quantities of labour and material, together with project preliminaries. Document likely to be some 200 pages in content, for which the CQS takes full responsibility
Contribute to preparation of list of tenderers	15	

Activity	Man-hours	Comment
Prepare, issue, receive and comment on tenderers' pre-qualification document	100	This activity usually involves a number of tenderers' interviews
Recommend form of contract, draft any amended clauses and complete specific project particulars	40	CQS usually heavily involved with contractual advice, often in conjunction with promoter's lawyers
Issue tender documents, receive tenders, analyse and recommend acceptance	100	CQS takes considerable responsibility for any recommendations made
Contingency	355	
Estimated man-hours up to contract award	3000	
Post-contract		
Draw up contract document for signing by parties	20	
Update cash flow forecast in line with successful tender	40	
Attend site meetings	75	
Prepare interim valuations for payment (monthly)	600	CQS responsible for advising architect but takes responsibility
Prepare cost report and update cash flow forecast	400	Developers will act on projected final cost and targeted draw-downs
Attend project meetings	200	
Value proposed changes prior to issue to contractor	75	
Agree value of changes with contractor	800	
Agree final account with contractor	200	
Issue final account to promoter including clear audit trail against expenditure	75	CQS takes responsibility for the accurate agreement of the final account and for its being auditable

Activity	Man-hours	Comment
Analysing, reporting on and negotiating contractor's claims under the contract	Not included	This can be a very intensive and important element of CQSs' activities. Usually remunerated on an hourly charge basis
Contingency	315	
Estimated man-hours following contract award	2800	

Pre-contract   3,000 hrs
Post-contract   2,800 hrs
Total   5,800 hrs
@ £35/hr = £203,000 (ECU 253,750)

This represents 2.03% of the construction budget (excluding land costs, finance costs, statutory and professional fees). (N.B. Incidental expenses are now usually included in the fee.)

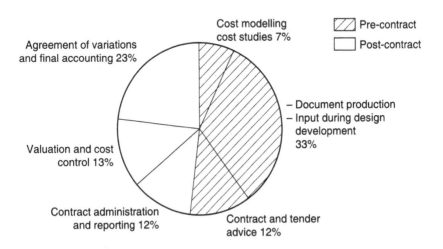

Fig. 5.2 Approximate apportionment of CQS activity during development cycle. Pre-contract activity ±52%; post-contract activity ±48%.

This illustrates a likely first assessment of a fee proposal for the project as defined. Should the CQS be in competition, a view will be taken as to how the bid should be pitched. It is likely that a range of between 1.5% and 2.0% would be achieved by the building promoter if between five and six firms are asked to tender. It is interesting to note that when compared against the RICS fully inclusive scale, the fee that would be charged represents 2.44% or £244,000 (ECU 305,000). This shows the effects of fee competition in a recessed construction market (see Appendix).

I hope it will be seen how British CQSs now approach their fee agreements. I do not doubt that the same procedure occurs in many other countries; what I believe is different, however, is the role and responsibilities that we, as construction economists, fulfil in our various construction industries.

The CEEC launched a 'desk study' into the comparison of construction economists' fee scales. This has proved impossible to do to date, for the reasons covered in this chapter. Most countries do, however, have some form of printed fee scale for either the individual function of the construction economist as a separate consultant, or as a construction economist working within the offices of another discipline (e.g. architect, engineer or project manager).

APPENDIX: RICS FEE SCALE 36 (1988 EDITION)

SCALE No. 36
INCLUSIVE SCALE OF PROFESSIONAL
CHARGES FOR QUANTITY SURVEYING
SERVICES FOR BUILDING WORKS

(29 July 1988)

1.0   GENERALLY

    1.1   This scale is for use when an inclusive scale of professional charges is considered to be appropriate by mutual agreement between the employer and the quantity surveyor.

    1.2   This scale does not apply to civil engineering works, housing schemes financed by local authorities and the Housing Corporation and housing improvement work for which separate scales of fees have been published.

1.3 The fees cover quantity surveying services as may be required in connection with a building project irrespective of the type of contract from initial appointment to final certification of the contractor's account such as:

(*a*) Budget estimating; cost planning and advice on tendering procedures and contract arrangements.

(*b*) Preparing tendering documents for main contract and specialist sub-contracts; examining tenders received and reporting thereon or negotiating tenders and pricing with a selected contractor and/or sub-contractors.

(*c*) Preparing recommendations for interim payments on account to the contractor; preparing periodic assessments of anticipated final cost and reporting thereon; measuring work and adjusting variations in accordance with the terms of the contract and preparing final account, pricing same and agreeing totals with the contractor.

(*d*) Providing a reasonable number of copies of bills of quantities and other documents; normal travelling and other expenses. Additional copies of documents, abnormal travelling and other expenses (e.g. in remote areas or overseas) and the provision of checkers on site shall be charged in addition by prior arrangement with the employer.

1.4 If any of the materials used in the works are supplied by the employer or charged at a preferential rate, then the actual or estimated market value thereof shall be included in the amounts upon which fees are to be calculated.

1.5 If the quantity surveyor incurs additional costs due to exceptional delays in building operations or any other cause beyond the control of the quantity surveyor then the fees shall be adjusted by agreement between the employer and the quantity surveyor to cover the reimbursement of these additional costs.

1.6 The fees and charges are in all cases exclusive of value added tax which will be applied in accordance with legislation.

1.7 Copyright in bills of quantities and other documents prepared by the quantity surveyor is reserved to the quantity surveyor.

2.0 INCLUSIVE SCALE

2.1 The fees for the services outlined in paragraph 1.3, subject to the provision of paragraph 2.2, shall be as follows:

(*a*)

<table>
<tr><td colspan="2">CATEGORY A:
Relatively complex works and/or works with little or no repetition.

Examples:
Ambulance and fire stations; banks; cinemas; clubs; computer buildings; council offices; crematoria; fitting out of existing buildings; homes for the elderly; hospitals and nursing homes; laboratories; law courts; libraries; 'one off' houses; petrol stations; places of religious worship; police stations; public houses, licensed premises; restaurants; sheltered housing; sports pavilions; theatres; town halls; universities, polytechnics and colleges of further education (other than halls of residence and hostels); and the like.</td></tr>
<tr><td>Value of Work</td><td>Category A Fee</td></tr>
<tr><td>£</td><td>£   £</td></tr>
<tr><td>Up to   150,000</td><td>380+6.0% (minimum fee £3,380)</td></tr>
<tr><td>150,000– 300,000</td><td>9,380+5.0% on balance over 150,000</td></tr>
<tr><td>300,000– 600,000</td><td>16,880+4.3% on balance over 300,000</td></tr>
<tr><td>600,000– 1,500,000</td><td>29,780+3.4% on balance over 600,000</td></tr>
<tr><td>1,500,000– 3,000,000</td><td>60,380+3.0% on balance over 1,500,000</td></tr>
<tr><td>3,000,000– 6,000,000</td><td>105,380+2.8% on balance over 3,000,000</td></tr>
<tr><td>over 6,000,000</td><td>189,380+2.4% on balance over 6,000,000</td></tr>
</table>

(*b*)

<table>
<tr><td>CATEGORY B:
Less complex works and/or works with some element of repetition.

Examples:
Adult education facilities; canteens; church halls; community centres; departmental stores; enclosed sports stadia and swimming baths; halls of residence; hostels; motels; offices other than those included in Categories A and C; railway stations; recreation and leisure centres; residential hotels; schools; self-contained flats and maisonettes; shops and shopping centres; supermarkets and hypermarkets; telephone exchanges; and the like.</td></tr>
</table>

Value of Work	Category B Fee
£	£ £
Up to 150,000	380+5.8% (minimum fee £3,260)
150,000– 300,000	9,060+4.7% on balance over 150,000
300,000– 600,000	16,110+3.9% on balance over 300,000
600,000– 1,500,000	27,810+2.8% on balance over 600,000
1,500,000– 3,000,000	53,010+2.6% on balance over 1,500,000
3,000,000– 6,000,000	92,010+2.4% on balance over 3,000,000
over 6,000,000	164,010+2.0% on balance over 6,000,000

(c)

CATEGORY C:
Simple works and/or works with a substantial element of repetition.

Examples:
Factories; garages; multi-storey car parks; open-air sports stadia; structural shell offices not fitted out; warehouses; workshops; and the ·like.

Value of Work	Category C Fee
£	£ £
Up to 150,000	300+4.9% (minimum fee £2,750)
150,000– 300,000	7,650+4.1% on balance over 150,000
300,000– 600,000	13,800+3.3% on balance over 300,000
600,000– 1,500,000	23,700+2.5% on balance over 600,000
1,500,000– 3,000,000	46,200+2.2% on balance over 1,500,000
3,000,000– 6,000,000	79,200+2.0% on balance over 3,000,000
over 6,000,000	139,200+1.6% on balance over 6,000,000

(d) Fees shall be calculated upon the total of the final account for the whole of the work including all nominated sub-contractors' and nominated suppliers' accounts. When work normally included in a building contract is the subject of a separate contract for which the quantity surveyor has not been paid fees under any other clause hereof, the value of such work shall be included in the amount upon which fees are charged.

(e) When a contract comprises buildings which fall into more than one category, the fee shall be calculated as follows:

(i) The amount upon which fees are chargeable shall be allocated to the categories of work applicable and the amounts so allocated expressed as percentages of the total amount upon which fees are chargeable.

(ii)　Fees shall then be calculated for each category on the total amount upon which fees are chargeable.

(iii)　The fee chargeable shall then be calculated by applying the percentages of work in each category to the appropriate total fee and adding the resultant amounts.

(iv)　A consolidated percentage fee applicable to the total value of the work may be charged by prior agreement between the employer and the quantity surveyor. Such a percentage shall be based on this scale and on the estimated cost of the various categories of work and calculated in accordance with the principles stated above.

(*f*)　When a project is the subject of a number of contracts then, for the purpose of calculating fees, the values of such contracts shall not be aggregated but each contract shall be taken separately and the scale of charges (paragraphs 2.1(a) to (e)) applied as appropriate.

2.2　*Air Conditioning, Heating, Ventilating and Electrical Services*

(*a*)　When the services outlined in paragraph 1.3 are provided by the quantity surveyor for the air conditioning, heating, ventilating and electrical services there shall be a fee for these services in addition to the fee calculated in accordance with paragraph 2.1 as follows:

Value of Work		Additional Fee	
£		£	£
Up to	120,000	5.0%	
120,000–	240,000	6,000+4.7% on balance over	120,000
240,000–	480,000	11,640+4.0% on balance over	240,000
480,000–	750,000	21,240+3.6% on balance over	480,000
750,000–	1,000,000	30,960+3.0% on balance over	750,000
1,000,000–	4,000,000	38,460+2.7% on balance over	1,000,000
	over 4,000,000	119,460+2.4% on balance over	4,000,000

(*b*)　The value of such services, whether the subject of separate tenders or not, shall be aggregated and the total value of work so obtained used for the purpose of calculating the additional fee chargeable in accordance with paragraph (a). (Except that when more than one firm of consulting engineers is engaged on the design of these services, the separate values for which each such firm is responsible shall be aggregated and the additional fees charged shall be calculated independently on each such total value so obtained.)

(c)   Fees shall be calculated upon the basis of the account for the whole of the air conditioning, heating, ventilating and electrical services for which bills of quantities and final accounts have been prepared by the quantity surveyor.

2.3   *Works of Alteration*
On works of alteration or repair, or on those sections of the work which are mainly works of alteration or repair, there shall be a fee of 1.0% in addition to the fee calculated in accordance with paragraphs 2.1 and 2.2.

2.4   *Works of Redecoration and Associated Minor Repairs*
On works of redecoration and associated minor repairs, there shall be a fee of 1.5% in addition to the fee calculated in accordance with paragraphs 2.1 and 2.2.

2.5   *Generally*
If the works are substantially varied at any stage or if the quantity surveyor is involved in an excessive amount of abortive work, then the fees shall be adjusted by agreement between the employer and the quantity surveyor.

3.0   ADDITIONAL SERVICES

3.1   For additional services not normally necessary, such as those arising as a result of the termination of a contract before completion, liquidation, fire damage to the buildings, services in connection with arbitration, litigation and investigation of the validity of contractors' claims, services in connection with taxation matters and all similar services where the employer specifically instructs the quantity surveyor, the charges shall be in accordance with paragraph 4.0 below.

4.0   TIME CHARGES

4.1   (a)   For consultancy and other services performed by a principal, a fee by arrangement according to the circumstances including the professional status and qualifications of the quantity surveyor.
       (b)   When a principal does work which would normally be done by a member of staff, the charge shall be calculated as paragraph 4.2 below.

4.2   (a)   For services by a member of staff, the charges for which are to be based on the time involved, such charges shall be calculated on the hourly cost of the individual involved plus 145%.

(b)   A member of staff shall include a principal doing work normally done by an employee (as paragraph 4.1 (b) above), technical and supporting staff, but shall exclude secretarial staff or staff engaged upon general administration.

(c)   For the purpose of paragraph 4.2 (b) above, a principal's time shall be taken at the rate applicable to a senior assistant in the firm.

(d)   The supervisory duties of a principal shall be deemed to be included in the addition of 145% as paragraph 4.2(a) above and shall not be charged separately.

(e)   The hourly cost to the employer shall be calculated by taking the sum of the annual cost of the member of staff of:

(i)   Salary and bonus but excluding expenses;

(ii)   Employer's contributions payable under any Pension and Life Assurance Schemes;

(iii)   Employer's contributions made under the National Insurance Acts, the Redundancy Payments Act and any other payments made in respect of the employee by virtue of any statutory requirements; and

(iv)   Any other payments or benefits made or granted by the employer in pursuance of the terms of employment of the member of staff.

and dividing by 1650.

5.0   INSTALMENT PAYMENTS

5.1   In the absence of agreement to the contrary, fees shall be paid by instalments as follows:

(a)   Upon acceptance by the employer of a tender for the works, one half of the fee calculated on the amount of the accepted tender.

(b)   The balance by instalments at intervals to be agreed between the date of the first certificate and one month after final certification of the contractor's account.

5.2   (a)   In the event of no tender being accepted, one half of the fee shall be paid within three months of completion of the tender documents. The fee shall be calculated upon the basis of the lowest original bona fide tender received. In the event of no tender being received, the fee shall be calculated upon a reasonable valuation of the works based upon the tender documents.
Note. In the foregoing context 'bona fide tender' shall be deemed to mean a tender submitted in good faith without major errors of computation and not subsequently withdrawn by the tenderer.

(*b*)   In the event of the project being abandoned at any stage other than those covered by the foregoing, the proportion of fee payable shall be by agreement between the employer and the quantity surveyor.

6

Methods of measurement: Community comparisons

Gerry O'Loughlin

6.1 INTRODUCTION

I agreed to undertake a study of the various methods of measurement in the different European countries on behalf of the Society of Chartered Surveyors in Ireland and the Comité Européen des Economistes de la Construction (CEEC), with the cooperation of link persons in other countries. The study was carried out in two stages. Stage 1 involved the preparation of a questionnaire to determine what methods of measurement actually exist in each country. Stage 2 involved a detailed examination of each method and its comparative complexity.

The questionnaire was circulated to contact people in the following countries:

- Belgium;
- Denmark;
- Finland;
- Germany;
- Ireland;
- The Netherlands;
- Portugal;
- Spain;
- Switzerland;
- United Kingdom.

The aim of the study was to determine the method of measurement in existence in each European country. In order to achieve this, it was necessary to:

- determine whether or not a method of measurement existed within each European country and whether or not it was nationally recognized;
- ascertain how widespread was its use, and which were the various bodies that contributed to its formation, i.e. government/institutions;
- describe the layout and contents of each method and the complexity of measurement;
- carry out a detailed examination of each method and its comparative complexity;
- determine the amount of information required from designers;
- based on the findings of this study, make recommendation on whether or not it was feasible and of benefit to develop a method of measurement suitable for the European construction industry.

6.2 HISTORICAL BACKGROUND TO THE STANDARD METHOD OF MEASUREMENT

The Standard Method of Measurement forms the basis for the measurement of the bulk of construction work in Ireland and the UK. The first edition was issued in 1922, with the expressed objective of providing a uniform method of measurement. Subsequent editions were issued as follows:

- First Edition 1922;
- Second Edition 1927;
- Third Edition 1935;
- Fourth Edition 1948;
- Fifth Edition 1963; amended 1964; metric 1968;
- Sixth Edition 1979;
- Seventh Edition 1988;
- Principles of Measurement International (with amendments) 1992.

6.2.1 Change within the industry

The primary objective of providing a uniform method of measurement has not changed with time. However, periodic developments and new editions

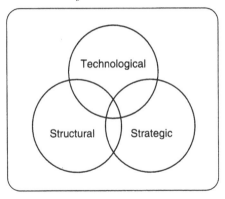

Figure 6.1 The main areas of change.

of the Standard Method of Measurement have occurred in response to the dynamic nature and changing needs of the construction industry. The main areas of change have been technological, structural and strategic (Fig. 6.1).

(a) Technological change

The evolution and transfer of technology has been manifest in the construction industry in a number of ways, through:

- the increased use of sophisticated service installations;
- the impact of information technology on the design and simulation of buildings;
- improvements in materials science, allowing for higher performance in structures and components;
- design incorporating buildability (design and build).

(b) Structural change

The fundamental structural change has been the enormous growth in subcontracting, with the complementary decline in main contracting in the

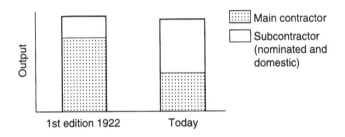

Figure 6.2 Structural change

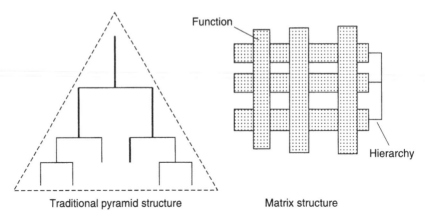

Traditional pyramid structure   Matrix structure

Figure 6.3 Organizational structure

actual physical construction process (Fig. 6.2). A significant number of subcontractors are now far bigger than some main contractors and, in many cases, more geographically spread. This structure must be reflected in the way we build, or else conflict will be inevitable.

The traditional way in which a project was structured relied on clear identification of roles, and hierarchical power. The organization was largely in a pyramid format, whereas today a matrix structure may be more applicable (Fig. 6.3).

(c) Strategic change

Changes in client requirements have resulted in the development of more focused project-delivery systems and treatment of fixed capital resources such as buildings. Many clients blithely suppose that it is possible to construct a top-quality building product at breakneck speed for a known-down price. The less rosy reality is that although the three most important considerations for a client are usually cost, time and quality, the business of procurement invariably calls for some compromise or conscious balancing of these priorities. The emergence of the management path for the procurement of construction projects is in part a measure of the influence of overseas investors – particularly from Japan and the USA – who have challenged traditional approaches.

The procurement systems evaluation sheet shown in Fig. 6.4 highlights the different procurement paths currently being used by clients. These range from the traditional approach, through design and build, to separate management functions of management contracting, construction management and project management, highlighting the different priorities of time, cost, quality, design, management, and specialist services. The system shown in this figure is used to arrive at the most effective type of procurement path for the successful completion of the project.

Contractors have responded to this highly competitive and complex market by developing both strategic policies and tactical plans in marketing, bidding, subletting and profit-maximizing their 'business operations'. Most contractors would tend to view themselves not just as construction companies but more as companies in construction.

Figure 6.5 illustrates an example of the strategic and tactical ploys that some construction companies now evaluate in tendering for projects. This example is based on the use of probability analysis with decision trees.

6.2.2 So where is this need for measurement?

Measurement takes place throughout a project life cycle and for many other reasons than just for preparing tender documentation (Table 6.1).

6.3 SUMMARY OF FINDINGS

The Standard Method of Measurement of Building Works (SMM 6th Edition) formed the basis for the measurement of the bulk of the construction work in Ireland at the time of the study, comparisons were, therefore made with SMM6. The results of this research indicated that seven of the eleven European countries use a method of measurement similar to SMM6. In most cases, these methods of measurement are nationally recognized.

- The layout of each method is generally in trade format, which is usually skills-based (carpentry, plastering etc.).
- The preambles section, defining the specification for materials and workmanship, is normally prepared by the architect.
- The preliminaries section, which identifies the particulars of the project in terms of the general facilities and obligations of the contractor, will very often form a separate document.
- In most cases the method is nationally recognized, but is not necessarily mandatory.

EVALUATIVE CRITERIA

Clients Priority:-		TRADITIONAL				DESIGN & BUILD						SEPERATE MANAGEMENT FUNCTION					
		Sequential		Accelerated		Negotiated		Competitive		Dev & Const.		Management Contrcting		Construction Management		Project Management	
Critical 5 / Desired 4,3 / Indifferent 2,1	Priority	Rating	Total	Rating	Total	Rating	Total	Rating	Total	Rating	Total	Rating	Total	Rating	Total	Rating	Total
TIME																	
Fastest Project Procurement	4	2	8	5	20	9	36	7	28	9	36	7	28	7	28	8	32
Critical Occupancy Date	5	3	15	3	15	8	40	8	40	8	40	8	40	6	30	7	35
Sectional Completion	1	5	5	5	5	8	8	7	7	8	8	8	8	7	7	7	7
COST																	
Fixed Priced	5	7	35	6	30	9	45	8	40	8	40	8	40	7	35	7	35
Lowest Capital Cost	2	9	18	7	14	9	18	10	20	8	16	8	16	6	12	5	10
Optium Life Cycle Cost	3	6	18	5	15	7	21	8	24	7	21	7	21	8	24	9	27
Funding Stream Certainty	4	7	28	6	24	9	36	9	36	8	32	7	28	6	24	7	28
Equity Participation by Contractor	1	6	6	7	7	9	9	9	9	10	10	9	9	5	5	6	6
DESIGN																	
Post Contract Flexibility	4	4	16	3	12	2	8	2	8	3	12	7	28	8	32	8	32
High-Tech, Specialised Design	4	5	20	5	20	5	20	5	20	6	24	6	24	8	32	9	36
QUALITY																	
Continous Performance Auditing	5	10	50	9	45	6	30	5	25	6	30	6	30	7	35	7	35
Complete Quality Certification	2	5	10	5	10	8	16	8	16	6	12	3	6	4	8	6	12
MANAGEMENT																	
Single Point Responsibility	5	4	20	4	20	10	50	10	50	8	40	6	30	8	40	9	45
Direct Professional Responsibility	4	8	32	8	32	4	16	4	16	4	16	9	36	9	36	10	40
SPECIALIST SERVICES																	
Tax Effeciency	3	1	3	1	3	4	12	5	15	6	18	3	9	6	18	8	24
Value Managed	4	2	8	2	8	8	32	8	32	7	28	6	24	4	16	9	36
Other			0		0		0		0		0		0		0		0
TOTAL-WEIGHTED FACTORS			292		280		397		386		383		377		382		440
COST (ESTIMATED) RELATIVE FACTOR			2.80		2.85		2.70		2.65		2.80		2.80		2.85		2.80
WEIGHT FACTOR COST RATIO			104.3		98.25		147		145.7		136.8		134.6		134		157

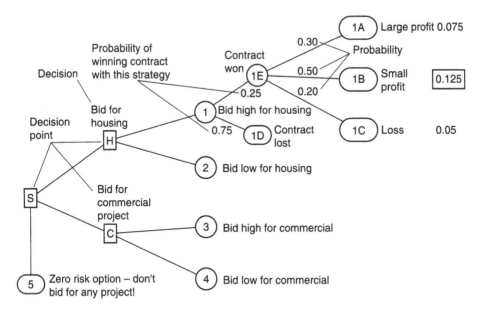

Fig. 6.5 Tendering strategy: decision and probability tree for analysing bidding tactics in housing and commercial projects.

- Each method is widely used, both in the private and public sectors, and will normally be the basis for the preparation of bills of quantities.
- Generally, each method is updated every ten years to cater for structural, technological and strategic change within construction and the greater economic environment. In the Netherlands it is updated every five years to achieve better coordination.
- Government bodies and professional institutions contribute to its formation in most of the countries.

Table 6.2 highlights the results of the questionnaire.

6.4 CONCLUSIONS

The absence of a uniform method of measurement in Europe provides a major stumbling block to the operation of a pan-European construction industry. Project costs and resource information is available, but at present is mainly contained in trade documentation, which limits its potential for

Table 6.1

Phase	Function
Inception stage	Develop feasibility and conceptual budgets. Decision to proceed or not
Design audit stage	Cost planning and control of design evolution
Contractor selection	Bidding document facilitating competitive tendering
Project planning stage	Assessing resource requirements and programming activities
Client funding stream	By producing physical resources and unit costs with the activity programme, an accurate profile can be simulated to reflect the client's financial obligation to the project
Variation and settlement	Reconciliation of work deviations. The more defined the original contract the greater the accuracy in valuation of change
Construction audit stage	Evaluating project performance and value against original agreements
Cost analysis/database	Comparative value study with previous projects and benchmarking for future projects

analysis. Data are not of a uniform nature and do not facilitate easy comparison between building types or regional locations or analysis for use as historic data. The bill of quantities, which is a major product of a standard method of measurement, has proved to be the single greatest source of historical cost information in Ireland, and has been instrumental in the creation of what is probably the largest construction databank in the world.

A method of measurement has many functions to fulfil in the project life cycle. The degree of complexity and detail to which construction works can be commonly measured to facilitate these functions would contribute greatly to providing an environment for competition by members of the European construction industry. As the CEEC strives to harmonize working methods between countries there is a need to assemble a database of cost information in a format accessible to all member states, which would culminate in the preparation of a European Method of Measurement.

The development of this document necessitates the establishment of the units of measurement in common use together with the grouping of construction work into functional elements. The estimating techniques in the EC are generally based on less complex measurements than in Ireland; this, therefore, suggests a simpler, more general method of measurement. However, it would not be strategically prudent to develop an over-simplified document for financial management of the project, from the point of view of both client and contractor.

Such a method of measurement must, however, recognize and reflect in its structure that the delivery processes involved in construction can be very different, even for the same type of construction, when different types of procurement systems or national customs are employed. These different approaches must not be restricted or compelled to produce unnecessary levels of measurement; on the contrary, the method of measurement must positively facilitate the pitching of the level of measurement necessary to the requirements of the procurement system. However, the research and compilation of any new method of measurement is a long, complicated and costly procedure, as can be borne out by the time taken for the development of previous methods.

The Department of the Built Environment, School of Engineering, RTC Limerick, has formulated proposals to research the extent to which members of the European construction industry understand the function of a method of measurement to:

- structure data in a format accessible to the European construction industry;
- facilitate a common method of measurement and value appraisal of construction at the various phases of a project life cycle;
- formalize a tendering procedure that optimizes competition of European construction companies.

One possible framework for establishing the above objectives is to use the Principles of Measurement International (POMI). The first edition of POMI was issued by The Royal Institution of Chartered Surveyors in the UK in 1979. Following some amendments it was adopted by the Society of Chartered Surveyors and the Construction Industry Federation in Ireland on a trial basis in April 1992. POMI could provide a focus for establishing existing and potential commonality among CEEC member countries.

The advantages of adopting POMI as the basis for a European Method of Measurement include the following:

- POMI was devised to facilitate the measurement of construction in an international context.
- Its brevity allows considerable room for expansion in detailed requirements should the need arise.

Table 6.2 Questionnaire results

Country	Method of measurement similar to SMM6		Name of method	Nationally recognized		Mandatory		Contractually required		Use			BOQs		Existence (years)	Update (years)	Why update?	Contribution
	Yes	No		Yes	No	Yes	No	Yes	No	Private	Public	Other	Yes	No				
Belgium	✓		Code de Mesurage	✓			✓	✓		✓	✓	✓		✓	15	10	New construction methods and technical progress	
Denmark		✓											✓					
UK	✓		SMM7	✓			✓	✓		✓	✓	✓	✓		66	10	Introduce new working, remove anachronistic work	
Finland	✓		Building 80	✓			✓	✓		✓	✓	✓	✓		20	10	Changes in building technology and estimating methods	Building 80 group
France	✓		Mode de Metre Normalise - Avant metre et Devis Quantitatif des Ouvrages de Batiment	✓			✓	✓		✓	✓		✓		20	10	New technique	UNTEC
Germany	✓		VOB/C - Verdingungsordnung fur Bauleistungen–Part C	✓		✓*		✓*		✓	✓	✓	✓		65	When necessary	For adoption	Government and societies of contractors
Ireland	✓		SMM6	✓			✓	✓		✓	✓	✓	✓		66	10	New techniques	RICS NFBTE and CF
Netherlands	✓		Nen 3699 Standard measurement of net quantities for materials and activities of building	✓			✓	✓		✓	✓	✓	✓		15	5	Better coordination	Various bodies, architects, engineers, quantity surveyors, contractors, subcontractors
Portugal		✓																
Spain		✓																
Switzerland		✓																

*For public contracts

- It has been used internationally since 1979 and as such has recorded an actual perspective of international measurement.
- It is currently available in English, German, French and Spanish.
- Its availability would allow the proposed research to be expedited.

As the CEEC strives to harmonize working methods between member states, so as to allow cost information to be exchanged between member countries and assemble a library of cost information in a format accessible to all members, then this committee should strive to achieve concord between member states, which would culminate in the preparation of a European Method of Measurement.

7

A sample case study of building projects

Bent Christensen

7.1 INTRODUCTION

This chapter presents the framework, the objectives and the visions for carrying out comparative studies for the realization or procurement of building projects in the EC member states from inception to occupation.

The objectives and the visions are:

1. to carry out comparative studies for the realization or procurement of projects in the member countries from inception to occupation;
2. to carry out sample case studies related to the same categories of buildings as to their functional use in the member countries;
3. to obtain experience and impressions about the price levels and professional practice in the member countries;
4. to study law and practice in the member countries and, possibly, make the application common;
5. to motivate the remaining member countries of the EC (Belgium, Greece, Italy and Luxembourg) and perhaps other European countries to participate in the study.

The sample case study of building projects is carried out by nationally elected members of the CEEC, who are responsible for the collection of data related to construction cost, professional fees, timescales and running costs. At present, the study has participation from eight EEC member states and Finland. The study comprises six different categories of buildings as to their functional use (school, administration, housing, factory, sport and senior citizens' housing) varying from 3,000 to 10,000 m² gross floor area. The study is partly completed and partly still going on.

7.2 THE MODEL

My proposal for a task-oriented approach is based on an ordinary functional diagram, a so called **box ball diagram**, with the application presented in Table 7.1.

Table 7.1 Model for carrying out a sample case study

Input	*Process*	*Output*
	Select drawings	Drawings
Drawings	Prepare bills of quantities	Bills of quantities
Bills of quantities	Pricing bills of quantities	Cost analysis Fee calculation Global costs
Drawings and specifications	Prepare timescales	Timescales

The model shows how the data from the sample cases are processed on the basis of selected drawings representative of the six categories of building projects mentioned in the introduction. Essential documents, such as bills of quantities, brief descriptions, guidelines and questionnaires, have been prepared centrally under my control.

The resulting analysis of construction costs, timescales, professional fees, global costs etc. will be made available to members and their national organizations upon completion for the purpose of early advice to clients.

7.3 ORGANIZING THE SAMPLE CASE STUDY

The rest of this chapter shows how the model may be used advantageously in the organization and management of sample case studies of building projects.

It is possible to structure experience from completed projects in the same way, for the exchange of project information among members: for example, through a national databank. The model of the sample case study illustrated in Table 7.1 consists of the following processes.

7.3.1 Selecting representative drawings

The categories and approximate sizes of the building projects to be ana-lysed had been defined at an earlier stage of the CEEC's work in the fol-lowing general terms:

1. a secondary school of approx 500 pupils;
2. an office block of 5,000–10,000 m² gross floor area;
3. a block of 50–100 local authority flats, 60–70 m² per flat, four storeys high with lift
4. an industrial building (e.g. factory), approx 3,000 m² with some ancil-lary office space;
5. a multi-purpose sports hall;
6. an old people's home of approx 50 units.

The idea was then that members of the Committee would be looking for suitable projects of the types described, which would form the basis for preparing the various documents, questionnaires etc. for each national team to complete. These preparatory documents have been produced centrally by myself with the assistance of enthusiastic colleagues.

7.3.2 Preparing bills of quantities

The most complicated and difficult process has been, and still is, the prepa-ration of the bills of quantities for the six projects in question, owing to variations in estimating methods and different needs for specification.

The bills of quantities prepared by me are so-called **elemental bills**, containing all relevant SfB-classified elements described as items for pric-ing. The idea is then that the national team that is doing the pricing should study how each individual element in that particular category of building traditionally is composed, and then price that composition.

An example taken from a housing project will show the principles of the idea. A typical item of the elemental bill of quantities, chosen at random, will appear as follows:

(21)Z.006 External wall construction

together with the quantity and possible dimensions if there are any differ-ences within that class. Further specification cannot be made centrally. The

SfB1	2	REF	TEXT	UNIT	QTY.	RATE	TOTAL

CONSTRUCTION ECONOMICS EUROPEAN COMMITTEE

SAMPLE CASE STUDY COMPLETED BY GERMANY

SCHOOL BUILDING

CURRENCY: INDEX: EXCL. VAT

CEEC5   3 APR 91

	SfB1	2	REF	TEXT	UNIT	QTY.	RATE	TOTAL
1	(14)	C	207	Excavation edge beam	m	26		
1	(14)	E		STAIR FOUND: CONCRETE WORKS				
1	(14)	E	004	In situ concrete edge beam	m	26		
2	(21)	A		EXT WALLS: PRELIMINARIES				
2	(21)	A	209	Prefabrication components	item			
2	(21)	Z		EXT WALLS: WORKS				
2	(21)	Z	006	External wall construction	m2	1715		
2	(22)	A		INT WALLS: PRELIMINARIES				
2	(22)	A	209	Prefabrication components	item			
2	(22)	F		INT WALLS: BLOCK WORKS				
2	(22)	F	007	Hollow wall	m2	116		
2	(22)	F	008	100 mm blockwork	m2	1253		
2	(22)	F	053	200 mm blockeork	m2	237		
2	(23)	E		FLOORS: CONCRETE WORKS				
2	(23)	E	054	150 mm reinforced concrete suspended slab	m2	1178		
2	(24)	E		STAIRS: CONCRETE WORKS				
2	(24)	E	057	Single storey dog-leg concrete stair	No.	3		
2	(24)	E	614	Single storey internal stair	No.	1		
2	(27)	A		ROOFS: PRELIMINARIES				
2	(27)	A	209	Prefabrication components	item			
2	(27)	E		ROOFS: CONCRETE WORKS				

Fig. 7.1

national pricing team must do the rest, otherwise you will probably write a specification of an element which is not relevant in the majority of countries participating in the study.

The same element is traditionally composed in three different countries as follows:

Denmark:

(21)Z.006 Building of 350 mm hollow wall of 110 mm red clay bricks, externally finished as the work proceeds, and 100 mm light-weight concrete elements, internally finished to receive wall-paper. The cavity is filled with 125 mm mineral wool.

United Kingdom:

(21)Z.006 External wall construction of dense aggregate concrete blocks (7 N/mm²) Forticrete 'Leicester Common' blocks or similar; in cement mortar (1:3). 100 mm thick in skins of hollow walls (50 mm cavity), solid.

Finland:

(21)Z.006 Cutting, assembly and erection of timber wall of 50 × 10 mm per 600 in both directions. Internal cladding of 13 mm gypsum board on and including dampproof membrane and 22 × 100 mm per 300 mm. External cladding of 25 × 100/25 × 130 mm vertically mounted on and including distance profiles and windproof membrane. The cavity of the wall is filled with 200 mm mineral wool.

It will be seen just from these three examples that national traditions, climate and the availability of resources influence the composition of functional elements in an interesting way, which hopefully will be a future task for a comparative study.

This sample case study is concentrated on the cost of an element, e.g. (21) External walls, but with the national specifics taken into account when pricing the bills of quantities.

Figure 7.1 is an example of the **elemental bill** prepared centrally by me and presented to the members/participants of the study; Fig. 7.2 is an example of a national interpretation of category (1), illustrating national traditions of elemental compositions in one category of building project in one member state.

7.3.3 Pricing bills of quantities

The basic information for pricing the bills of quantities is of course this document itself, together with a few representative drawings for visual

	SfB1	2	REF	TEXT	UNIT	QTY.	RATE	TOTAL
				CONSTRUCTION ECONOMICS EUROPEAN COMMITTEE				
				SAMPLE CASE STUDY COMPLETED BY GREAT BRITAIN				
				SCHOOL BUILDING				
				CURRENCY: INDEX: EXCL. VAT				
				CEEC6 3rd April 1991				
1	(14)	C		STAIR FOUND: EARTH WORKS				
1	(14)	C	207	Excavation edge beam 400 x 600 mm X-Section area. Excavation for pik caps and ground beams between piles; n.e. 1 m deep.	m	26		
1	(14)	E		STAIR FOUND: CONCRETE WORKS				
1	(14)	E	004	In situ concrete edge beam 400 x 600 mm X-Section area. Reinforced insitu concrete grade C20, 20 mm aggregate (1:2:4) to ground beams	m	26		
2	(21)	A		EXT WALLS: PRELIMINARIES				
2	(21)	A	209	Prefabrication components 50 mm cavity; galvanised steel butterfly wall ties @ 3/m2	item	1		
2	(21)	K	210	Cavity wall insulation; 50 mm "Drieherm" cavity wall filling	item	1		
2	(21)	Z		EXT WALLS: WORKS				
2	(21)	Z	006	External wall construction Dense aggregate concrete blocks (7N/mm2) Forticrete "Leicester Common" blocks or similar; in cement mortar (1:3). 100 mm thick in skins of hollow walls (50 mm cavity), solid.	m2	1715		
2	(22)	A		INT WALLS: PRELIMINARIES				
2	(22)	A	209	Prefabrication components	item			
2	(22)	F		INT WALLS: BLOCK WORKS				

Fig. 7.2

impression and for evaluation of the work's complexity. An A4 page of general specification is provided as well.

The majority of members participating in the study have no problems in connection with pricing of items regarding construction cost, but some construction economists are not familiar with calculation of costs involving other disciplines such as bank and financial experts, architects and engineers, and must therefore establish communication lines with these skills in order to reach meaningful results.

Figure 7.3 is an example of a priced bill of quantities, and Figs 7.4 and 7.5 illustrates the resulting analysis for one category of building project from one country.

The CEEC Concordance for the Exchange of Cost Data shown in Fig. 7.4 is the result of the work of the former Cost Commission of CEEC, whereas the analysis shown in Fig. 7.5 is the automatic result of applying the international SfB classification system for computer processing of bills of quantities and documents/analysis related thereto.

7.3.4 Preparation of timescales

By April 1992, the former Timescale Commission had completed the comparative study of detailed timescale information for the planning and construction of the six projects mentioned in the introduction of this chapter. The comparative study comprises the durations of 16 main activities, from initial planning and design, through tender documentation, to construction and completion.

Figures 7.6, 7.7 and 7.8 show examples of the results of the comparative study of the above-mentioned subject, documented in a report available from the member organizations.

7.4 CONCLUSION

During the different phases of the design and construction process, individual clients require different ranges of services from their consultants: for example, practical advice on the likely cost of the project, recommended optimum quality of materials and method of construction; the length of the construction period; means of keeping the project within budget and on schedule; management of the project to provide a single link between the rest of the planning and construction team. These are some of the functions undertaken by the construction economist. It is therefore important to facilitate the exchange of experience and project information between professionally qualified persons who are responsible for construction economics in the EC member states. One way is to participate in a sample case study as described in this chapter.

```
CONSTRUCTION ECONOMICS EUROPEAN COMMITTEE

SAMPLE CASE STUDY COMPLETED BY GERMANY

SCHOOL BUILDING

CURRENCY: DM                   INDEX: 1.010            EXCL. VAT (14%)
```

SfB1	2	REF	TEXT	UNIT	QTY.	RATE	TOTAL

CEEC1 — 3 APR 91

	SfB1	2	REF	TEXT	UNIT	QTY.	RATE	TOTAL
1	(13)	C		FLOOR FOUND: EARTH WORKS				
1	(13)	C	001	150 mm harcore	m2	2166	15	32490
1	(13)	E		FLOOR FOUND: CONCRETE WORKS				
1	(13)	E	002	150 mm reinforced concrete slab	m2	2166	50	108300
1	(13)							140790
1	(14)	C		STAIR FOUND: EARTH WORKS				
1	(14)	C	207	Excavation edge beam	m	26	15	390
1	(14)	E		STAIR FOUND: CONCRETE WORKS				
1	(14)	E	004	In situ concrete edge beam	m	26	85	2210
1	(14)							2600
2	(21)	A		EXT WALLS: PRELIMINARIES				
2	(21)	A	209	Prefabrication components	item			0
2	(21)	Z		EXT WALLS: WORKS				
2	(21)	Z	006	External wall construction	m2	1715	550	943250
2	(21)							943250
2	(22)	A		INT WALLS: PRELIMINARIES				
2	(22)	A	209	Prefabrication components	item			0
2	(22)	F		INT WALLS: BLOCK WORKS				
2	(22)	F	007	Hollow wall	m2	116	75	8700

Fig. 7.3

Secondary School 3440 m2   CEEC Concordance for the Exchange of Cost Data

Country of Origin   DK

Currency   DKK

Rates for items 1 to 29 exclude preliminaries

which add 13.1 % to cost

	Total Cost of Element	Cost per sq.m GFA	% of total Cost
1) SUBSTRUCTURE	1,097,000	319	4.3
SUPERSTRUCTURE			
2) Frame	1,488,000	433	5.8
3) External walls	214,000	62	0.8
4) Internal walls	657,000	190	2.6
5) Floors	406,000	118	1.6
6) Roofs	726,000	211	2.8
7) Stairs	124,000	36	0.5
8) Windows and external doors	1,753,000	510	6.8
9) Internal doors	682,000	198	2.7
Group element total	6,050,000	1,758	23.6
FINISHES			
10) Internal wall finishes	884,000	257	3.4
11) External wall finishes	100,000	29	0.4
12) Floor finishes	1,146,000	333	4.5
13) Ceiling finishes	858,000	249	3.3
Group element total	2,988,000	868	11.6
14) EQUIPMENT AND FURNISHINGS	3,582,000	1,041	13.8
SERVICES			
15) Plumbing	193,000	56	0.8
16) Heating	784,000	228	3.0
17) Ventilating and Air Conditioning	2,085,000	606	8.1
18) Internal drainage	112,000	33	0.4
19) Electrics	1,408,000	409	5.4
20) Communications	268,000	78	1.0
21) Lifts, Escalators etc.	355,000	103	1.4
22) Protective installations	258,000	75	1.0
23) Miscellaneous services installation	30,000	9	0.1
Group element total	5,493,000	1,597	21.2
Sub total excluding External Works, Preliminaries, and Contingencies			
EXTERNAL SITE WORKS			
24) Site preparation	210,000	61	0.8
25) Site enclosure	216,000	63	0.8
26) Site fittings	358,000	104	1.4
27) Site services	957,000	278	3.7
28) Site buildings	278,000	81	1.1
29) Hard and soft landscaping	1,187,000	345	4.6
Group element total	3,206,000	932	12.4
SUB-TOTAL excluding Preliminaries and Contingencies	22,416,000	6,515	86.9
30) PRELIMINARIES incl. elem. not specified	3,354,000	975	13.1
TOTAL EXCL. CONTINGENCIES	25,770,000	7,490	100.0

Fig. 7.4

```
CONSTRUCTION ECONOMICS EUROPEAN COMMITTEE  - EXCHANGING COST DATA
FIRM              : Mangor & Nagel A/S, DK, - Tel. 45-42 31 32 10
LOCATION          : Frederikssund DK CATEGORY       : School
BUILDING TYPE     : New construction NO. OF FLOORS  : 2
CLIENT CATEGORY   : Public              INDEX       : R 115
FORM OF CONSULTANCY : Overall service FEE PER CENT  : 13.2
START OF CONSTRUCTION: July 1986     CONSTRUCTION TIME: 18 months
SITE AREA         : 11,816 m2           BUILT AREA  : 2166 m2
GROSS FLOOR AREA  : 3440 m2             BASEMENT AREA : 148 m2
```

| SCHOOL | | | | | | 10 OCTOBER 1991 |

B	E	ELEMENT	KR/M2	T.KR 1000	% OF C.C.	T.KR 1000	% OF C.C.	DESCRIPTION
		ELEMENT DESCRIPTION						Secondary school building of approx 500 pupils. The designs have been prepared by the French delegation of the CEEC for the purpose of exchanging e.g. cost data. In this connexion, it should be imagined that the project at one time was built in the town of Frederikssund, approx. 40 km north west of Copenhagen, and all the cost data were registered as it will be seen hereinafter.
		SITE						
	1	Site substructure	61	210	0.8			Excavating and levelling of topsoil
	2	Site buildings	81	278	1.1			
	3	Site enclosure	63	216	0.8			
	4	Site finishes	244	839	3.3			
	5	Site services	226	777	3.0			Surface water drainage and foul drainage
	6	Site installations	52	180	0.7			Supply cable and external lighting installations
	7	Site fittings	104	358	1.4			
	8	Landscaping	101	348	1.3			
			-----	-----	-----			
			932	3206	12.4			
			-----	-----	-----			
1		SUBSTRUCTURE						
1	1	Excavation	18	62	0.2			
1	2	Foundations	82	283	1.1			
1	3	Floor beds	214	736	2.9			
1	4	Basem. stairs, ramps	5	16	0.1			
1	5							
1	6	Building drainage						
1	7	Piling						
1	8	Substructure, others						
-			-----	-----	-----			
1			319	1097	4.3			
-			-----	-----	-----			
2		SUPERSTRUCTURE						
2	1	External walls	62	214	0.8			
2	2	Internal walls	190	657	2.6			

Fig. 7.5

```
CONSTRUCTION ECONOMICS EUROPEAN COMMITTEE  - EXCHANGING COST DATA
FIRM            : Mangor & Nagel A/S, DK, - Tel. 45-42 31 32 10
LOCATION        : Frederikssund DK CATEGORY       : School
BUILDING TYPE   : New construction NO. OF FLOORS  : 2
CLIENT CATEGORY : Public             INDEX         : R 115
FORM OF CONSULTANCY : Overall service FEE PER CENT  : 13.2
START OF CONSTRUCTION: July 1986     CONSTRUCTION TIME: 18 months
SITE AREA       : 11,816 m2          BUILT AREA    : 2166 m2
GROSS FLOOR AREA : 3440 m2           BASEMENT AREA : 148 m2
```

SCHOOL					10 OCTOBER 1991

B	E	ELEMENT	KR/M2	T.KR 1000	% OF C.C.	T.KR 1000	% OF C.C.	DESCRIPTION
2	3	Floor slabs	118	406	1.6			
2	4	Stairs	36	124	0.5			
2	5							
2	6	Balconies						
2	7	Roofs	211	726	2.8			Roof construction only, roofing and flashings see 47
2	8	Superstruct., others	433	1488	5.8			Frame
-			-----	-----	-----			
2			1050	3615	14.1			
-			-----	-----	-----			
3		COMPLETIONS						
3	1	External wall compl.	510	1753	6.8			Windows and door/window-elements
3	2	Internal wall compl.	198	682	2.7			Internal doors
3	3	Floor screeds						
3	4	Banisters	11	39	0.2			
3	5	Suspended ceilings						
3	6	Balconies, compl.						
3	7	Rooflights	113	388	1.5			
3	8	Completions, others						
-			-----	-----	-----			
3			832	2862	11.2			
-			-----	-----	-----			
4		FINISHES						
4	1	Ext. wall finishes	29	100	0.4			Painting of wood work externally
4	2	Int. wall finishes	257	884	3.4			Wall tiling and painting on glass fibrous fabric
4	3	Floor finishes	333	1146	4.5			Painted concrete, terrazzo, ceramic tiling and linoleum
4	4	Stair finishes	19	65	0.3			
4	5	Ceiling finishes	249	858	3.3			
4	6	Balcony finishes						
4	7	Roof finishes	565	1944	7.5			
4	8	Finishes, others	7	23	0.1			Painting of plumbing installations
-			-----	-----	-----			

Side 2

Fig. 7.5 (continued)

```
CONSTRUCTION ECONOMICS EUROPEAN COMMITTEE  -  EXCHANGING COST DATA
FIRM              : Mangor & Nagel A/S, DK, - Tel. 45-42 31 32 10
LOCATION          : Frederikssund DK CATEGORY        : School
BUILDING TYPE     : New construction NO. OF FLOORS   : 2
CLIENT CATEGORY   : Public            INDEX          : R 115
FORM OF CONSULTANCY : Overall service FEE PER CENT   : 13.2
START OF CONSTRUCTION: July 1986      CONSTRUCTION TIME: 18 months
SITE AREA         : 11,816 m2         BUILT AREA     : 2166 m2
GROSS FLOOR AREA  : 3440 m2           BASEMENT AREA  : 148 m2
```

| SCHOOL | | | | | | 10 OCTOBER 1991 |

B	E	ELEMENT	KR/M2	T.KR 1000	% OF C.C.	T.KR 1000	% OF C.C.	DESCRIPTION
4			1459	5020	19.5			
-			-----	-----	-----			
5		SERVICES						
5	1	Refuse disposal	9	30	0.1			
5	2	Drainage	33	112	0.4			
5	3	Water services	56	193	0.8			
5	4	Gas services	40	138	0.5			
5	5	Cooling services						
5	6	Heating services	188	646	2.5			
5	7	Ventilation/air cond	606	2085	8.1			General ventilation incl. recirculation and exhaust sporadically
5	8	Services, others						
-			-----	-----	-----			
5			932	3204	12.4			
-			-----	-----	-----			
6		ELECTRICAL INSTALLAT						
6	1	Electrical centre	39	135	0.5			
6	2	Power installations	105	361	1.4			
6	3	Lighting installat.	265	912	3.5			
6	4	Communication	78	268	1.0			
6	5							
6	6	Lifts	103	355	1.4			
6	7	Transport installat.						
6	8	Electr. inst.,others	75	258	1.0			Fire alarm and burglar alarm installations
-			-----	-----	-----			
6			665	2289	8.8			
-			-----	-----	-----			
7		FITTINGS						
7	1	Reception fittings	53	182	0.7			
7	2	Working room fit.ngs	569	1958	7.6			Including classroom furniture
7	3	Kitchen fittings	196	674	2.6			
7	4	Toilet room fittings	158	544	2.1			

Side 3

Fig. 7.5 (continued)

```
CONSTRUCTION ECONOMICS EUROPEAN COMMITTEE  - EXCHANGING COST DATA
FIRM                     : Mangor & Nagel A/S, DK, - Tel. 45-42 31 32 10
LOCATION                 : Frederikssund DK CATEGORY     : School
BUILDING TYPE            : New construction NO. OF FLOORS : 2
CLIENT CATEGORY          : Public           INDEX        : R 115
FORM OF CONSULTANCY      : Overall service FEE PER CENT  : 13.2
START OF CONSTRUCTION: July 1986        CONSTRUCTION TIME: 18 months
SITE AREA                : 11,816 m2        BUILT AREA   : 2166 m2
GROSS FLOOR AREA         : 3440 m2          BASEMENT AREA : 148 m2
```

SCHOOL						10 OCTOBER 1991

B	E	ELEMENT	KR/M2	T.KR 1000	% OF C.C.	T.KR 1000	% OF C.C.	DESCRIPTION
7	5	Maintenance fittings	2	7	0.0			
7	6	Store room fittings	3	10	0.0			
7	7							
7	8	Miscellaneous	60	207	0.8			
-			-----	-----	-----			
7			1041	3582	13.8			
-			-----	-----	-----			
8		TEMPORARY WORKS						
8	1	Site facilities	174	600	2.3			
8	2	Running of site	42	145	0.6			
8	3	Climatic precautions	44	150	0.6			
8	8	Other facilities						
-			-----	-----	-----			
8			260	895	3.5			
-			-----	-----	-----			
9		PROFESSIONAL FEES						
9	1	Architect				1614	6.3	Including design management
9	2	Structural engineer				747	2.9	
9	3	Mechanical engineer				688	2.7	
9	4	Electrical engineer				341	1.3	
9	5	Constr. economist				250	1.0	
9	6	Site management				860	3.3	
9	7	Land surveyor				250	1.0	
9	8	Other costs				742	2.9	
-						-----	-----	
9						5492	21.4	
-			-----	-----	-----	-----	-----	
T			7490	25770	100.0	5492	21.4	
-			-----	-----	-----	-----	-----	

Side 4

Fig. 7.5 (continued)

Detailed time scale information. Summary of sample case studies.

Project: Secondary school of approximately 500 pupils

Duration: w = weeks, m = months, y = years

Country: Operation:	D	DK	E	F	GB	IRL	NL	P	SF
1. Appoint design team	2 w	2 w	1 w	2 w	4 w	2 w	3 w	4 w	7 w
2. Establish brief	8 w	12 w	1 w	4 w	8 w	8 w	4 w	3 w	7 w
3. Pre–contract design:									
– outline proposal	8 w	6 w	1 w	6 w	10 w	3 w	3 w	6 w	7 w
– scheme design	8 w	8 w	6 w	8 w	8 w	6 w	5 w	12 w	21 w
– detail design	12 w	8 w	10 w	8 w	8 w	6 w	7 w	10 w	14 w
4. Cost control and –verification:									
– cost estimate	2 w	1 w	1 w	2 w	2 w	1 w	1 w	3 w	3 w
– cost plan	4 w	4 w	6 w	3 w	2 w	2 w	2 w	6 w	3 w
– cost verification	2 w	2 w	8 w	3 w	6 w	1 w	1 w	6 w	2 w
5. Authority controls:									
– building permission	6 m	8 w	4 w	9 w	16 w	12 w	8 w	8 w	3 w
6. Site development	–	–	4 w	–	–	–		–	–
7. Client: go–ahead!	2 w	8 w	1 w	3 w	10 w	6 w	4 w	4 w	4 w
8. Tender documents:									
– specification, BQ, instructions	12 w	3 w	4 w	2 w	8 w	10 w	4 w	–	4 w
9. Tender period	6 w	3 w	4 w	4 w	4 w	4 w	3 w	5 w	5 w
10. Analysis of tenders	4 w	2 w	1 w	2 w	1 w	2 w	2 w	3 w	3 w
11. Select/appoint contractors	3 w	3 w	1 w	2 w	2 w	1 w	2 w	3 w	3 w
12. Mobilization period	3 w	4 w	4 w	4 w	2 w	2 w	3 w	4 w	–
Planning period, total	13 m	12 m	9 m	11 m	12 m	11 m	10 m	13 m	12 m
13. Construction period	24 m	18 m	12 m	18 m	18 m	15 m	14 m	15 m	12 m
14. Making good defects	2 m	2 m	4 w	2 m	6 m	6 w	2 m	2 m	4 w
Construction period, total	26 m	20 m	13 m	20 m	18 m	15 m	16 m	15 m	13 m
15. Defects liability period	5 y	5 y	1 y	1 y	12 y	1 y	10 y	1 y	1 y
16. Final payment/disengagement	1 y	6 m	4 w	4 w	3 m	4 w	3 m	4 w	6 w

Fig. 7.6

Graphical Summary of planning and construction periods. 1 month = 0.8 mm

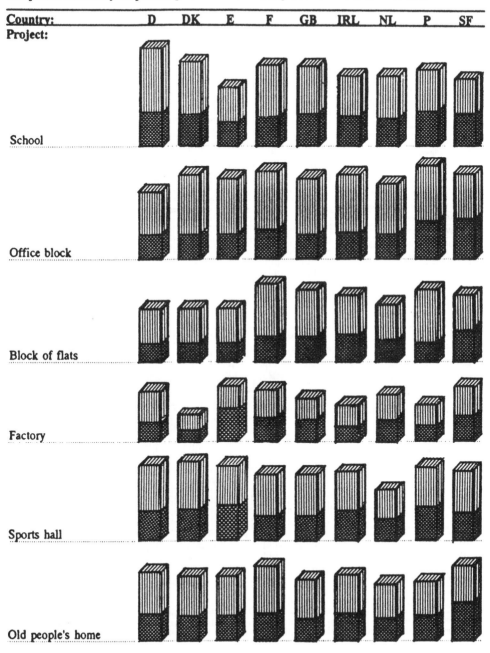

Fig. 7.7

BAU: schule - 500 schulern
Bundesrepublik Deutschland

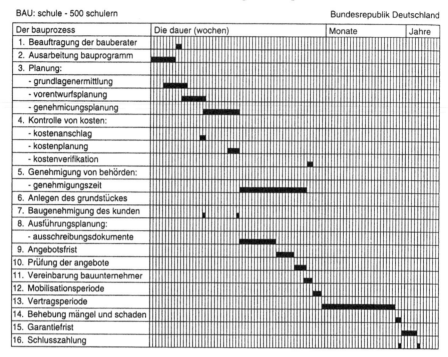

Der bauprozess	Die dauer (wochen)	Monate	Jahre
1. Beauftragung der bauberater			
2. Ausarbeitung bauprogramm			
3. Planung:			
- grundlagenermittlung			
- vorentwurfsplanung			
- genehmicungsplanung			
4. Kontrolle von kosten:			
- kostenanschlag			
- kostenplanung			
- kostenverifikation			
5. Genehmigung von behörden:			
- genehmigungszeit			
6. Anlegen des grundstückes			
7. Baugenehmigung des kunden			
8. Ausführungsplanung:			
- ausschreibungsdokumente			
9. Angebotsfrist			
10. Prüfung der angebote			
11. Vereinbarung bauunternehmer			
12. Mobilisationsperiode			
13. Vertragsperiode			
14. Behebung mängel und schaden			
15. Garantiefrist			
16. Schlusszahlung			

Building type: secondary school of approx 500 pupils
Denmark

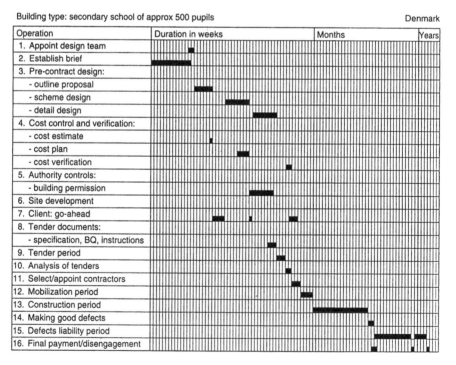

Operation	Duration in weeks	Months	Years
1. Appoint design team			
2. Establish brief			
3. Pre-contract design:			
- outline proposal			
- scheme design			
- detail design			
4. Cost control and verification:			
- cost estimate			
- cost plan			
- cost verification			
5. Authority controls:			
- building permission			
6. Site development			
7. Client: go-ahead			
8. Tender documents:			
- specification, BQ, instructions			
9. Tender period			
10. Analysis of tenders			
11. Select/appoint contractors			
12. Mobilization period			
13. Construction period			
14. Making good defects			
15. Defects liability period			
16. Final payment/disengagement			

Fig. 7.8

8

Building as an investment process

John Connaughton

8.1 INTRODUCTION

The capital investment decisions of firms account for a good deal of the construction output of the developed world, and yet very little consideration has been given to the corporate client's role in the building process. The now extensive literature on building economics, procurement and management has very little to say about the role of clients, for whose benefit the entire construction process takes place. In particular, this literature embodies a vision of the client as a single individual who is decisive, certain of his building requirements and clear in his articulation of these to the construction industry.

I shall argue that while this might apply to some of the industry's clients, it by no means applies to all of them. I am particularly concerned with the large corporate client that invests in a new building for its own use as a production facility, but much of what I have to say applies equally to the large organization that requires a new building to further some organizational or business objective. Although an examination of why firms and organizations need new buildings would provide fruitful topics for discussion, I am more concerned here with *how* firms obtain new buildings as part of a capital investment decision.

8.2 WHO ARE OUR CLIENTS?

As construction professionals we have a strong interest in who our clients are and what they want. Many of today's construction clients are large corporations with worldwide operations that – in terms of annual value of sales compared to gross national product – are larger than many nation states. Their capital investment decisions should be of considerable interest to us and, in particular, how they go about the process of obtaining suitable buildings for their own uses.

But let us first address the question: 'Who are our clients?': who within corporate organizations makes the decision to invest in new buildings? The theory of capital investment lays great stress on the identification of investment opportunities and on gathering and evaluating information so that investment choices will be made that will maximize return of investment (Bierman and Smidt, 1988). The point is that the problem is treated as one of **choice**. And such choices are made by top management.

Indeed, the 'theory' of building procurement and management – if there is one – tends to characterize corporate investors in a similar manner. This literature assumes that the corporate client – represented by its senior management, who have taken the decision to invest in new building work – can act as a decisive and authoritative individual. Consider, for example, the literature on briefing (Kelly *et al.*, 1992). In the UK there is now a considerable body of guidance on 'good practice' in the construction briefing process, much of which is written from a construction industry perspective. This emphasizes the role of the client as providing information to facilitate the process of building design. Where corporate clients – or client organizations – are considered, they are required to be represented by a senior, authoritative and decisive individual who can articulate 'his' requirements clearly and with certainty.

Within the literature on construction management also, relatively little attention is paid to the corporate client or to the client as an organization. The emphasis here is on forms of project organization, and on management to time and cost criteria (Walker, 1989). Indeed, the dominant theme in much of this literature is that of project leadership, which tends to be placed in the hands of professional project managers. The client's role is in providing information and approving decisions. And in this, clients have much in common with their counterparts in the briefing literature: they are required to act as a unitary entity, to be clear, decisive and authoritative.

However, recent work in the business administration area of social science, in challenging the characterization of the investment process within theories of corporate finance, raises further questions about the corporate client's role in the construction process. Empirical studies of the process of capital investment in the large firm present a view of the investment process that is radically different from that presented by finance theory. Whereas the latter considers the process of investment to involve top management in choices between investment alternatives for the allocation of scarce investment resources, empirical research has shown investment to be more of a 'bottom-up' process (Marsh *et al.*, 1988). Here, investment proposals originate from deep within the operating divisions of large corporations and progress up the management hierarchy until they emerge, fully packaged as proposals for funding, before top management. Top management then either accept or reject these proposals, but do little to change them. Indeed, top management participate very little in the formulation and development of these proposals in the first place.

What are the implications – if any – of this picture of corporate capital investment for current construction practice, which, I have argued, emphasizes the unitary nature of the corporate client? First, these studies describe investment as an organizational and political process as much as an economic or financial decision. The key participants in the process – those who determine what new buildings are required – are located deep within the corporate organization. The emphasis in the construction literature that the corporate client be represented by a senior authoritative individual may be misplaced.

Second, as will be seen, the process by which firms decide what new buildings they require occurs over long periods of time and across many levels of the corporate hierarchy. This involves the firm in an internal process of bargaining and choice, which is not reflected in the emphasis within the construction literature on the unitary client. Corporate clients are complex organizations and not the single-minded sole entrepreneurs of traditional economic theory.

I have recently undertaken two in-depth case studies of the process by which large firms made and implemented capital investment decisions, to examine these implications more closely. In particular, I wanted to examine how the large firms studied defined and obtained the new buildings they needed, given the reported 'bottom-up' nature of their investment decision processes. Furthermore, by looking at the construction process from the client's perspective and as part of a capital investment decision – and not as an end result of it – the study was intended to provide additional insights into the corporate client's role. To do this, a model of the process by which large firms made investment decisions was needed.

8.3 A MODEL OF INVESTMENT DECISION-MAKING

The 'bottom-up' process of corporate capital investment has been charted by a number of authors, including Bower (1970) in the USA and King (1975) in the UK. In the two case studies recently undertaken, I have used the model developed by Bower as a conceptual scheme. Bower's study highlighted the importance of capital investment as a process of resource allocation spread over long periods of time and over many levels of the corporation. He argued that, unlike the prescriptive theories of finance and economics, which emphasized the role of top management choice, investment proposals arose from the divisions. They were developed and attracted support as they progressed through the management hierarchy, arriving with top management as ready-packaged projects, which were either accepted (more likely) or rejected, but hardly ever changed.

He proposed a scheme for the examination of this process, which charts the identification and development of investment projects in terms of three processes: project definition, impetus and determination of context. These processes have different phases associated with them, which are broadly related to hierarchical levels within the firm: initiating, integrating and corporate (Table. 8.1).

Bower examined capital investment projects that led to the acquisition of new production facilities, and argued that the process by which capital investment projects were defined (**definition**) was initiated by lower-level managers within the divisions, whose concerns were production- or 'facility'-oriented. These managers identified a 'discrepancy' between the production needs of their business and their current capability. Between identifying such a discrepancy and submitting a proposal to top manage-

Table 8.1 Bower's model of resource allocation

	Process		
Phase	*Definition*	*Impetus*	*Determination of context*
Corporate	Aggregate, financial company – environmental	Yes or no	Design of corporate context
Integrating	Financial aggregate ↑↓ Product – market strategic	The company wants ↑↓ The businesses want	Corporate need ↑↓ Subunit need
Initiating	Strategic product – market	'I've got a great idea'	Product – market not served by structure

ment, the technical aspects of this definition were largely resolved at this level of the firm. The extent to which these potential investment projects moved towards funding depended on the support provided by middle-ranking managers (**impetus**). Top management did not participate directly in the process but could influence it by changing the organizational 'context' within which lower-level managers operated (this process Bower called **determination of context**). Bower's analysis owes much to the picture of man within organization developed by Simon (1976) and firms' internal decision processes outlined by, for example, Cyert and March (1963). An important element in Bower's model is the activity of **integration**, which is necessary to reconcile the needs of divisional managers with those of top management, and draws heavily on Lawrence and Lorsch (1967).

8.4 SOME QUESTIONS FOR RESEARCH

The process of capital investment described by Bower ends with approval of an investment proposal by top management. All that is left to do is to build what has been defined in the investment proposal and within the limit of funding approved by top management. But what happens when projects change? Let us consider this problem for a moment, because change following approval of the investment proposal provides an opportunity to use Bower's model during the implementation/construction phase to examine management action.

For the manufacturing firm seeking a new factory building for its own use as a production facility, conventional economic theory tells us that factory buildings are required as inputs to production. The demand for these buildings is therefore related to the demand for the products to be produced in them. Because the building process takes time, the firm must produce the level of productive capacity it will require when the new factory will finally be available; in fact, it is likely to take a longer-term view of production needs (Nutt, 1988) as new building investment arises relatively infrequently.

However, between ordering and receiving a new building, the firm's production requirements may change owing to changes in product demand, changes in technology, and so on. We can consider this change in terms of Bower's model to help explain the role of the firm's management in the construction process. Following the approval of the investment proposal, those managers who were involved in defining the building contained within it may be expected to monitor their requirements to help ensure that they get the building the firm needs. Indeed they must do this or accept that the definition contained in the proposal will constitute the sole definition of the project, even when requirements subsequently change.

The process of monitoring, assessing and, if necessary, changing project definition during implementation involves prediction, analysis and review or confirmation of project viability, and is essentially part of an ongoing capital investment process. More specifically, the process of redefining the project may be considered in terms of the process of definition in Bower's model. It will be triggered by a discrepancy between anticipated production needs and planned capacity. In the language of the model, the discrepancy will be identified by those whose concerns are 'facility-oriented' in response to information that indicates, for example, that sales forecasts were too low, or that a new product or production process is required. In other words, change to the construction project that is concerned with production capacity issues will arise from deep within the client organization.

Therefore, when the change needed is of such a nature that approval by top management is required, the proposal for change will need sponsorship at successively higher levels than where it originates. The process of getting projects changed may therefore be viewed in terms of the process of impetus in Bower's model. However, an additional, critical factor is present when the project is being implemented. As well as obtaining top management approval for such change, the participation and cooperation of construction firms must also be obtained. During implementation, then, impetus is about the firm's management of external as well as internal processes.

8.5 BUILDING AS AN INVESTMENT PROCESS

It is now possible to consider Bower's model as a framework for the examination of management action during project implementation (Table. 8.2). The processes of redefining the project and incorporating changes may be described in terms of the model's subprocesses of definition and impetus. The initiating phase is the primary determinant of redefinition: as with definition, a discrepancy identified by facility-oriented managers that triggers this process is expected to constitute the main source of project redefinition. The integrating phase is the primary determinant of the process of changing projects.

So how then do firms make and implement industrial building investment decisions? And how do they ensure that they get the buildings they need?

Table 8.2

Phase	Process		
	Definition (redefinition)	Impetus (implementing change)	Determination of context
Corporate	Aggregate, financial	Yes or no	Determine/agree structure
Integrating	Financial aggregate ↓ ↑ Product – market strategic	The company wants ↓ ↑ The businesses want	Corporate need ↓ ↑ Subunit need
Initiating	Product – market Will product market needs be served by the new building?	'I've got a great idea' 'The original idea needs to be changed'	Product – market not served by structure

To put the question another way, how does the firm implement a capital investment decision in conditions where its definition of its building requirements is uncertain? In the two projects I examined – one was a 2000 m² factory for the manufacture of antibiotics for a large UK-based pharmaceuticals company; the other a large 30,000 m² factory for the manufacture and assembly of engineering equipment – the 'bottom-up' pattern of investment decision-making was observed. Both proposals involved substantial investment (around £10 million each at 1993 prices). Proposals for investment originated in the operating divisions, and in one of the cases – the pharmaceuticals factory – it was almost two years before the proposal was formally submitted to the board for funding approval.

On both projects, proposals were moved towards funding by a process of 'impetus' whereby managers at successively higher levels lent support to the proposal. By the time both proposals arrived before their respective boards for approval, there had been a considerable level of managerial time and resource invested in them. On the pharmaceutical factor, managers below board level had already begun allocating expenditure within their control to enable construction work to commence before the main board approved the proposal. There was a sense in which the implementation process had already begun prior to the completion of the investment decision.

Furthermore, both proposals were approved quickly by their respective boards without query or referral. There was little involvement by top management in the investment decision processes; additionally, there was little attempt by top management to adjust the organizational structure to influence the decisions of those lower-level managers involved in the investment projects.

On both projects, construction consultants were involved early in the definition process – prior to submitting investment proposals to the main board – and this involvement was managed by those managers who had responsibility as construction client. The model provided a useful means of examining this, and the 'integrating phase' of the definition process helped to explain how both firms defined the buildings they needed. These managers had a considerable task of integrating the needs of the businesses (for additional productive capacity) with the needs of top management (for growth in corporate earnings) and both of these needs with questions of what could be built, when, where and at what cost. The process of definition was therefore managed by these 'integrating-level' managers, who carried out the client function.

On the pharmaceutical factory project, two major changes occurred – one shortly before the proposal was submitted to top management and the other after work had commenced on site. Both of these changes originated from those managers who had been involved in the definition of building needs, but both changes required considerable support from higher up the management hierarchy before they could be incorporated, primarily because of the impetus the project had already attracted by then. Impetus can help to move investment proposals towards funding, but it can also act as a force against major change.

8.6 SOME IMPLICATIONS FOR CONSTRUCTION PROFESSIONALS

The scope of this chapter has permitted only a very brief overview of the detailed casework and the theoretical background. However, an examination of how firms obtain the new buildings they need as part of the process of capital investment has provided interesting insights into the corporate client's role in construction projects. First, the firm's role in the definition of its requirements as observed is rather different from that portrayed in the briefing literature. There, the client's role is as provider of information on which a building design may be formulated. While this process is important, it tends to ignore the wider context within which major investment decisions of the kind observed are made. In particular, the manager acting as construction client will need to reconcile the production needs of his particular business (in the case of a factory building) with the top management need for corporate earnings. This complicates the extent to which that manager may be able to provide clear, authoritative, statements of the firm's building needs. Furthermore, the focus of the briefing literature on senior management may be misplaced – the real decisions about building needs are made deep within the corporate organization. Top management in large corporations participate very little – if at all – in the investment decision-making process. The building 'client' may not be – and probably does not need to be – a senior manager.

Second, there is little attention paid in the construction management literature in particular to the extent to which clients may need to monitor – and sometimes change – their requirements during the construction process. On both projects examined, the client had a central role in the definition process, which went far beyond that of provider of information and authorizer of decisions. This need to manage the definition process in particular to ensure that a suitable building is obtained requires methods of building procurement and management that accommodate this role and the potential for change during the construction process.

REFERENCES

Bierman, H. and Smidt, S. (1988) *The Capital Budgeting Decision: Economic Analysis of Investment Projects*, 7th edn, Macmillan, New York.

Bower, J.L. (1970) *Managing the Resource Allocation Process*, Harvard Business School Press, Boston.

Cyert, R.M. and March, J.G. (1963) *A Behavioral Theory of the Firm*, Reprinted 1970, Prentice-Hall, Englewood Cliffs, NJ.

Kelly, J., MacPherson, S. and Male, S. (1992) *The Briefing Process: A Review and Critique*, RICS Research Papers, Paper number 12, Royal Institution of Chartered Surveyors, London.

King, P. (1975) Is the emphasis of capital budgeting theory misplaced? *Journal of Business Finance and Accounting*, **2**(1), 69–82.

Lawrence, P.C. and Lorsch, J.W. (1967) *Organization And Environment: Managing Differentiation and Integration*, Harvard University Press, Boston, MA.

Marsh, P., Barwise, P., Thomas, L. and Wensley, R. (1988) *Managing Strategic Investment Decisions in Large Diversified Companies*, London Business School, Centre for Business Strategy, London.

Nutt, B. (1988) The strategic design of buildings. *Long Range Planning*, **21**(4), 130–140.

Simon, H.A. (1976) *Administrative Behavior: A Study of Decision Making Process in Administrative Organization*, 3rd edn, Macmillan, New York.
Walker, A. (1989) *Project Management in Construction*, 2nd edn, BSP Professional Books, Oxford.

9

La naissance d'une directive RRGA?

Rob de Wildt

Depuis la deuxième moitié des années quatre-vingts, la Commission de la Communauté Européenne étudie la possibilité d'une directive européenne pour la réception, la responsabilité, les garanties et les assurances dans le secteur de la construction. L'année 1988 a vu la publication d'une recherche comparative réalisée par la FIEC, l'organisme européen des entrepreneurs dans la construction. Une première étude dans ce domaine a été executée par M. Mathurin (1989) par order de la Commission. Pourtant son rapport offrait peu de points de départ pour la Commission Européenne d'entreprendre d'autres actions: puisqu'il se contentait surtout d'une présentation des advantages du système français. C'était de début d'un bon bout de chemin à faire pour arrive (oui ou non) à une directive concernant la réception, la responsabilité, les garanties et les assurances (RRGA) dans le secteur de la construction.

Un grand nombre de participants dans la construction ne pouvait pas apprécier les initiatives de la part de la CE. Les uns sont d'avis que l'harmonisation des règles sur un niveau européen crée une législation inutile: la construction est une affaire locale ou régionale. Les opérations internationales étant d'un nombre relativement restreint, chaque participant devrait lui-même se renseigner sur le pays vers lequel il oriente ses activités d'exportation. Les autres allèguent qu'une directive qui se limite aux thèmes mentionnés, n'aboutira qu'à un simulacre de harmonisation: à cause des différences fondamentales dans le secteur de la construction, la diversité dans les pratiques de la construction continuera à exister. Ainsi dans les Pays-Bas l'opinion est en vogue qu'une directive devrait s'étendre non seulement à l'RRGA, mais qu'elle devrait s'étendre également aux conditions générales administratives (ce qui pourrait être considéré comme un effort de faire trop pour arriver nulle part). La résistance la plus forte de la part du secteur de la construction (le côté de l'offre) vient néanmoins de ces pays où une directive européene aurait pour conséquence un agrandissement de la responsabilté par rapport à la situation actuelle (l'Allemagne). De l'autre côté on trouve les organisations des consommateurs c.q. les clients (le côté de la demande) dans les pays qui en ce moment connaissent une législation relativement étendue sur le terrain de la responsabilité (la France).

9.1 LA DIRECTIVE RESPONSABILITÉ PRESTATION DE SERVICES

Pendant une longue période ces opinions opposées pouvaient s'allier dans une léthargie commune. Cette situation changeait au moment où les officiels du DG III organisèrent une réunion au début de 1991 pour un total de plus de 20 organisations européennes dans le domaine de la construction, des clients jusqu'aux constructeurs, des architectes jusqu'aux approvisionneurs de matériaux de construction.

Peu de temps avant un projet de directive avait été publié relativement à la responsabilité en cas de prestation de services et élaboré sous la responsabilité du DG XI (affaires de consommateurs). Dans ce projet le secteur de la construction était chargé d'une responsabilité extrèmement lourde: pendant une période de 20 ans il serait question d'une responsabilité solidaire de tous les participants concernés. Chacun dans le secteur de la construction était convaincu que cette directive ne pouvait être appliquée (à l'exception des architectes néerlandais). DG III signalait que – si le secteur de la construction voulait échapper à un tel régime – la collaboration des participants dans se secteur serait indispensable pour arriver à une directive particulière pour la construction.

Le message était clair: étant donné que les tentatives antérieures aient échouées, le secteur de la construction elle-même devrait se mettre au travail; ce travail comprenant aussi bien la coordination des concertations et l'apport sur le plan du contenu, que le compte-rendu destiné à la Commission Européenne. En plus ce travail devrait dans un demi an (avril-octobre 1991) aboutir à un projet de texte que le DG III pourrait utiliser comme base pour un projet de directive. Seulement ainsi il deviendrait possible de séparer le secteur de la construction des autres secteurs de service contenus dans la directive générale pour la prestation de services.

9.2 GAIPEC

On a pu constater que la suite de cette réunion aussi était déjà prévue: La FIEC s'offrait comme coordinateur, et pour une très grande partie c'était M. Paetzold, manager de bureau à Bruxelles sur qui retombait le travail. En plus quatre groupes de travail étaient créés qui s'occupaient successivement de la réception, la responsabilité, les garanties et les assurances. Ce travail se faisait sous le titre de 'Groupes des Associations Interprofessionnelles Européennes de la Construction', ou bien GAIPEC. Comme représentant du CEEC et de l'AEEBC, j'avais l'honneur de pouvoir participer au groupe de travail 2, à savoir le groupe de responsabilité. Pour la plupart des membres de ces groupes de travail, les réunions étaient intéressantes et fécondes. Cette méthode de travail imposée par la CE, pourra se prêter à répétition sur d'autres terrains où un accord sur un niveau européen est à souhaiter. En principe le texte requis par le DG III était prêt fin 1991, début 1992. Les coûts de cette procédure étaient très bas pour la Commission, surtout si on les compare avec les coûts normaux pour les études étendues par des rapporteurs spéciaux.

Entretemps, un an et demi est passé et peu de choses ont eu lieu pendant cette période. Le texte était terminé en janvier 1992, certes, mais il durait jusque septembre 1992 pourque la FIEC présente ce texte à la Commission (1). D'après ce qu'on dit parce qu'un ou plusieurs membres de la FIEC s'opposent toujours avec force à une directive particulière. En attendant, un référendum avait été organisé au Danemark et on s'était efforcé de trouver une explication practicable du terme 'subsidiarité'. En tout cas cela signifiait pour la Commission Européenne qu'une publication à court terme d'un projet de directive RRGA ne serait point réalisable. En attendant un livre blanc sur l'harmonisation de la législation dans la CE, il serait mieux d'attendre aved d'autres démarches. Depuis la deuxième moitié de 1992 un grand nombre de participants montre beaucoup moins d'enthousiasme lorsqu'il s'agit de consacrer leur temps à l'unification européenne. Aujourd-hui un revirement pourrait se produire, si au moins 1% de la population danoise a changé d'opinion.

Avec cela un tableau de la situation a été brossé et nous pouvons continuer avec la deuxième moitié, le projet GAIPEC pour une directive.

9.3 DIRECTIVE GAIPEC

Le développement d'une directive de responsabilité pour la construction a été surtout une recherche du juste équilibre entre les dispositions qui sont actuellement en vigueur et qui sont très diverse, entre offre et demande, entre les désavantages évitables et les avantages difficiles à réaliser. Un participant bien vite en avait assez: le CECODHAS, l'organisation européenne des offices HLM, il se retirait bien vite des groupes de travail.

Le travail a été accéléré parce que les participants pouvaient assez vite accepter quelques principes de base. Le principe central est que chaque participant ne peut être responsable que pour son propre travail, conformément aux obligations légales et contractuelles, et doit aussi assumer cette responsabilité. Un système comme celui en France, où il est question d'une responsabilité civile de risques du constructeur pendant une période de dix ans, devra beign mener à une situation où il fault assurer tous les risques.

La limitation de la responsabilité de chaque participant aux propres activités implique qu'on assume une responsabilité de culpabilité et non une responsabilité de risques. La directive ne prenait pas en considération une responsabilité pénale. Le participant dans le processus de construction est tenu d'une façon de travailler méticuleuse et professionnelle. Pourtant cela veut dire qu'un client qui désure que la responsabilité de risques soit couverte, doit obtenir cela par une autre voie. La responsabilité de risques à couvrir se réfère d'une part aux risques que le client court, également en cas d'une méthode de travail méticuleuse et professionnelle et d'autre part aux erreurs possibles par rapport au chevauchement d'activités par les participants individuels.

Pour obtenir la couverture de la responsabilité des risques il existe deux possibilités: une extension des obligations contractuelles ou bien une assurance supplémentaire. Les divers participants dans le processus de la construction devraient eux-mêmes se charger de la responsabilité de culpabilité. L'assurance de cette responsabilité appartient au domaine de l'assurance de responsabilité professionelle. Il est difficile d'attendre des partenaires individuels la couverture de la responsabilité de risques à l'intérieur d'un réseau parfois tellement compliqué comme la préparation et la réalisation de la construction. C'est la raison pour laquelle une assurance est indispensable.

Les représantants du secteur de la construction en Europe ont dès le début reçu le mesage que la Commission Européenne elle-même déterminerait le délai de la responsabilité. Cela n'empêchait pas les membres du 'groupe de travail responsabilité' de conseiller un délai de 5 ans pour la réception. Les participants était unanimement d'accord, bien qu'il faille mentionner qu-entretemps le CECODHAS – dans sa qualité de représentant d'un grand groupe de clients – avait jeté l'éponge et ne participat plus. L'équilibre encontré entre le conseil à la Commission et la liberté explicite de politique prise par la Commission permettent d'éviter une longue polémique au sujet du délai de responsabilité.

Le rapport-GAIPEC comprend cinq sections:

- Définitions;
- Portée et cession de droits;
- Réception;
- Responsabilité;
- Garantie.

La portée est limitée aux immeubles résidentiels et non résidentiels, conformément à l'ordre de la Commission. La discussion a laissé entrevoir la possibilité d'une inclusion des construction (y compris les travaux de génie civil, c.q. les constructions de route et d'ouvrages hydrauliques), à l'exception de la garantie facultative dans ce cas. La responsabilité regarde tous les participants concernés de la construction.

9.4 RÉCEPTION

La responsabilité est liée à une date initiale: la réception. Celle-ci appartenait au domaine d'un groupe de travail séparé, qui prenait à tâche de trouver une réglementation univoque pour la réception. Ce groupe a conçu une seule réception, qui pourra avoir lieu sans ou avec des remarques. Au cas où la réception ait lieu avec des remarques, il faut en même temps – ou de préférence en avance – fixer un délai pour une réponse satisfaisante aux remarques. La réception pourra être sollicitée par une des deux parties, ou bien pourra être fixée par contrat.

La directive introduit un certificat de réception, dans lequel le client confirme par écrit d'accepter le bâtiment et consigne aussi les éventuelles remarques. Un tel certificat est valable sou réserve de ce qui ait été réglé au niveau national.

9.5 RESPONSABILITÉ

Comme indiqué ci-dessus le délai de responsabilité entre en vigueur directement après la réception. Si le projet de construction est compliqué, il y aura beaucoup de participants. Si le bâtiment montre une imperfection et il deviendra nécessaire de rendre responsables un ou plusieurs participants en fait de leurs activités, il sera également nécessaire de démontrer que le participant en question était responsable de cette imperfection. Il pourrait s'agir d'un participant autre que le celui qui a produit la partie en question: il est bien concevable que ce n'était pas le producteur mais l'architecte, l'archeteur ou le surveillant qui était en défaut. Bien qu'il ait été stipulé que chacun est responsable à titre individuel pour ses propres défauts, il est bien possible qu'il s'agit d'une responsabilité collective pour une imperfection déterminée. Il est indiqué aussi qu'il y a huit raisons pour la dispense de responsabilité, comme les dégâts provoqués par un tiers ou par la partie qui porte plainte. Les participants ne pourront pas être rendus responsables de dégâts si ceux-ci ce sont produits en conséquence d'innovations qui ont été sollicitées par le client, comme suite de défauts cachés en cas d'un bâtiment déjà existent, ou par manque de connaissance par rapport au comportement de certaines constructions (comme à l'époque la corrosion du béton). Le client doit informer le participant responsable des défauts dans un délai d'un an après que ces défauts se soient manifestés.

9.6 GARANTIE

Si le client/propriétaire désire couvrir le risque de ne pas pouvoir rendre responsable celui qui était à l'origine des défauts, alors une garantie sur l'édifice sera souhaitable. Une telle garantie pourra être offerte par un des participants, par un assureur ou un institut spécial de garantie. L'inconvénient le plus important contre un cautionnement par un des participants est que le garantie perd sa validité en cas de faillite ou de liquidation judiciaire du participant-garant.

La garantie a les fonctions suivantes:

- un prompt règlement des dégâts: il n'est pas nécessaire que le propriétaire désigne le ou les participants responsable(s), il n'a qu'à démontrer le défaut et les dommages qui en sont la conséquence. Le garant – doit s'adresser aux éventuels responsables.

- la couverture des dommages pour lesquels personne ne pourra être rendue responsable à cause de l'impossibilité de trouver un ou plus responsables. ici il s'agit des cas où le problème de la culpabilité ne peut pas être résolu de manière univoque.
- la couverture des dommages pour lesquels personne ne pourra plus être rendue responsable, par disparition du participant responsable.

Il est au client/propriétaire de se pourvoir d'une telle garantie s'il le désire. Le client a aussi le liberté de ne pas vouloir une telle garantie. Ses raisons pourraient être: la capacité de se charger lui-même des dégâts, la confiance dans la professionnalité des participants et plus en général d'estimer les coûts de la garantie supérieurs aux avantages. Le rapport du GAIPEC fait exception pour les logements: ceux-ci devront être pourvus d'une garantie. Ici on met sur le même plan – d'après mon avis injustement – les logements des propriétaires individuels et les corporations professionnelles de gestion de logements.

La garantie comprend une limite inférieure à fixer en détail, sous laquelle la garantie n'est pas obligatoire. A côté de cela il existe la possibilité de n'assurer que les dommagres qui dépassent un risque base déterminé. En principe la garantie court parallèlement au délai de responsabilité de (cinq) ans, mais il est concevable que le client/propriétaire accorde une garantie à plus long terme. Le garant aura le droit de rendre responsable les participants responsables.

9.7 CONSÉQUENCES POUR LE SECTEUR DE LA CONSTRUCTION

Dans chaque pays de la CE une directive comme esquissée ici donnera un nouveau visage au monde de la construction, avec le seule exception peut-être de la France, où pendant une période de dix ans de responsabilité a conduit à une interaction intensive entre le monde des assurances et la construction. Selon ceratins cette situation allait trop loin parce que par la responsabilité civile les coûts de construction comprenaient au fond un délai de garantie de dix ans. Dans la directive RRGA le client sera libre de choisir entre les deux possibilités.

Le propriétaire individuel d'un logement fait exception, il manque souvent d'expertise afin de pouvoir connaître les possibilités de se protéger contre les risques. Pour cette raison la garantie est obligatoire dans ce cas. Une telle garantie existe déjà en ce moment, entre autres aux Pays-Bas et en Angleterre. Aux Pays-Bas cette garantie est fournie par un institut de garantie, qui s'oriente en particulier vers l'acheteur individuel d'un logement. L'institut prend soin des condition de police uniformisées et du règlement des dégâts. La garantie elle-même est fournie par le constructeur et assurée par quelques fonds de garantie collectifs. En plus de la responsabilité directement après réception, le risque de faillite pendant la construction est coassuré. Cette garantie est obligatoire en cas d'un logement subventionné. Depuis quelques années la part des logements garantis prend du volume parmi les logements non subventionnés.

Une des questions qui est soulevée par le système qui est proposé ici est de savoir comment le monde des assurances jugera ce système. Pendant le travail du GAIPEC les assureurs étaient représentés, mais il est clair qu'ils n'étaient pas très enthousiastes à donner de l'information sur la practicabilité de la garantie proposée. A part de la société d'assurances, il existe la possibilité, indiquée ici, des fonds et instituts de garantie. La question sur l'acceptation de la garantie obligatoire est liée à ses coûts: les coûts pouvaient-ils rester restreints à 1 ou 2% des frais de constructions? Vu les expériences avec l'institut néederlandais de garantie on peut donner une réponse positive à cette réponse: là les coûts se montent de 0,5% à 1% des frais de construction. Ce pourcentage inclut une couverture quelque peu élargie parce aux Pays-Bas la responsabilité de l'architecte et de l'ingéniuer en cas ce projects majeurs va jusqu'au maximum de leurs honoraires.

Pourque la garantie reste payable, il faudra prendre soin qu'un niveau de qualité trop bas sera puni. Si les participants peuvent utiliser la garantie pour compenser un niveau de qualité inférieure, la garantie produit l'effet inverse. Les coûts de la garantie monteront très vite dans ce cas, les frais de construction étant transformés en frais de garantie. Le règlement des dégâts et défauts est toujours moins profitable que les bénéfices obtenus à court terme d'une qualité inférieure et une augmentation vertigineuse des frais en sera la conséquence. Ceci est une raison importante de s'adresser aux participants et de leur parler de leur responsabilité, tel qu'il est proposé dans le rapport du GAIPEC.

9.8 LA PRACTICABILITÉ DE LA DIRECTIVE RRGA

La directive européenne est bien esquissée dans le rapport du GAIPEC. Il reste pourtant à savoir si la Commission prendra en considération cette directive pendant les activités à venir. Le principe de subsidiarité pourra conduire à la mise à part de la directive. La Commission présente deux arguments pour faire une directive: la protection de consommateurs/clients et un marché ouvert. Ses idées en quesiton sont peu claires: la directive pour la prestation de services est développée en partant de l'idée de la protection des consommateurs/clients (DG XI), tandis qu'ici l'argument du marché ouvert pèse beaucoup plus dans la balance. Et en réaction à cela, ceux qui s'occupent du marché ouvert (DG III) sont en train de préparer une directive qui finalement s'oriente surtout vers la protection des consommateurs (garantie).

Si la protection des consommateurs l'emporte, alors le principe de la subsidiarité pèsera plus lourd: chaque pays doit lui-même donner une forme à cette protection, en fonction des règles de droit en vigueur dans ce pays et des procédures de construction habituelles du même pays. Par contre: le marché ouvert exige de travailler à l'échelle de la CE, ceci par définition. En ce moment on donne beaucoup moins d'importance au marché ouvert qu'il y a un ou deux ans. Et il reste un point d'interrogation si la directive RRGA forme un pas important vers le marché ouvert. Les règles de droit à la base de chaque pays et la habitudes dans le secteur de la construction qui s'y ajoutent, font que l'harmonisation de celle-ci n'est qu'un tout petit pas sur le chemin vers le marché ouvert. Mais ce pas, il faut bien le faire.

9.9 SIGNIFICATION POUR L'ECONOMISTE DE LA CONSTRUCTION

L'economiste de la construction doit bien être au courant des problèmes éventuels autour de la responsabilité et des garanties. S'il y aura beaucoup de changements sur ce terrain, il est important qu'il en prenne connaissance et en étudie les conséquences. C'est pour cette raison que les évolutions autour de la directive RRGA sont esquissées ici. Le résultat de ce processus d'évolution est encore difficile à prévoir. Reste encore à savoir s'il y a aussi des conséquences plus directes pour le travail de l'economiste.

Un changement dans les règles relativement à la responsabilité et aux garanties n'a qu'une influence limitée pour le travail de l'economiste. L'economiste de la construction doit en cas d'innovations permettant de réduire les coûts, se mettre bien au courant des effets sur la garantie. Il est bien concevable que ces réductions seront contrebalancées par les coûts plus élevés de la garantie. En plus, des propos à economiser peuvent inclure une responsabilité de l'economiste après réception. Dans l'avenir le domaine du règlement des dégâts aura une influence plus grande dans la construction. L'economiste de la construction a ici un terrain facile vers lequel il pourra s'orienter.

Les coûts de la garantie ou de l'assurance se répercutent de deux façons sur l'estimation de l'economiste de la construction: puisque tous les participants devront adapter leur responsabilité professionnelle, leur tarif pourra s'élever (ou baisser) et en outre les coûts de la garantie influeront sur les frais de construction. Bien que son imoprtance puisse rester restreinte à environ 1%, ce poste ne peut manquer dans un avant-métré solide.

10

Exchange of data

Keith Hudson

10.1 COST COMMISSION

The CEEC from its inception has sought to identify common problems and areas of concern and development in the profession of construction economics. This resulted in the formation of a number of Working Commissions, one of which was the Cost Commission. The Cost Commission was established with terms of reference to advise the Committee on cost data, to harmonize working methods to allow cost data to be exchanged, and to assemble a library of cost information.

The Cost Commission turned its attention to defining basic terms and then to the problems involved in exchanging data. It decided to concentrate on building costs. It has defined those costs as being the capital costs to the building client as tendered by the successful contractor and being exclusive of land purchase costs, professional fees for design and VAT. The primary unit for the exchange of cost data was agreed as the cost per m^2 of gross internal floor area: that is, the total of all enclosed floor space measured to the internal face of the enclosing walls, with no deduction for internal walls, partitions, stairwells or lift shafts.

10.2 CONCORDANCE DOCUMENT

From the discussion on the exchange of data it became clear that almost all member countries produced analysis of tender prices on an elemental basis (foundations, walls, roof, heating installation etc.). It transpired that there was a great similarity between the list of elements employed.

Country of Origin   _____   Rates for items 1 to 29 exclude preliminaries
Currency   _____   which add _____% to cost

	Total Cost of Element	Cost per m² GFA	% of total cost
1) SUBSTRUCTURE			
SUPERSTRUCTURE			
2) Frame			
3) External walls			
4) Internal walls			
5) Floors			
6) Roofs			
7) Stairs			
8) Windows and extl. doors			
9) Internal doors			
Group element total			
FINISHES			
10) Internal wall finishes			
11) External wall finishes			
12) Floor finishes			
13) Ceiling finishes			
Group element total			
14) EQUIPMENT & FURNISHINGS			
SERVICES			
15) Plumbing			
16) Heating			
17) Ventilation and Air Conditioning			
18) Internal drainage			
19) Electrics			
20) Communications			
21) Lifts, Escalators etc.			
22) Protective installations			
23) Miscellaneous services inst.			
Group element total			
Sub total excluding External Works, Preliminaries and Contingencies			
EXTERNAL SITE WORKS			
24) Site preparation			
25) Site enclosure			
26) Site fittings			
27) Site services			
28) Site buildings			
29) Hard and soft landscaping			
Group element total			
SUB-TOTAL excluding Preliminaries and Contingencies			
30) PRELIMINARIES			
TOTAL EXCL. CONTINGENCIES			

Fig. 10.1 CEEC Concordance for the exchange of cost data

Four options for harmonization were examined:

- the adoption of one of the existing member country forms of analysis;
- production of a European Standard;
- the compilation of conversion factors to convert analyses from one country to those of another;
- production of a 'concordance document' to which members' analyses could be readily converted.

It was felt that any concordance document should reflect the determinants of cost: quantity, quality and price.

It was agreed to produce a concordance document together with a conversion protocol for each member country. This protocol would identify which sub-elements of a member country's analysis were subsumed by each concordance element. A concordance document and conversion protocol has been produced for each of the existing member countries.

It would be relatively simple to write these protocols as computer programs, making conversion of national analyses to concordance format a simple matter. Conversion by computer opens up possibilities of a European database. This aspect is covered in greater depth by Douglas Robertson in Chapter 19.

The production of large numbers of concordance analyses would be expensive; therefore a shortened form of analysis, called the **Concise Concordance**, was produced, which will serve to index the projects on which full data are available (Fig. 10.1).

10.3 UNIT RATES

The concordance analyses record the cost per m² of projects disaggregated into construction elements and expressed in local currencies. In order to obtain project cost in one's own currency it is necessary to convert one currency to another. The conversion of project costs from one currency to another depend on the vagaries of monetary exchange rates. These are notoriously fickle. In fact, in the present currency market, if you converted the cost of a project from one currency to another in the morning you would probably be out of date by the afternoon of the same day.

Therefore it seemed a logical step to attempt a more precise comparison of prices between member countries, and so a list of some 63 items of building work in common use was compiled. These items were defined with some precision and some time was spent in removing ambiguities and ensuring that all members had a common understanding of the various items (e.g. one member thought that a two-panel door was a double door). It was agreed that prices should be for a four-storey office block of approximately 1,100 m² gross floor area, situated on the edge of a large city, in reinforced concrete framed construction, for a contract let on a fluctuating price basis, prices to include a reasonable allowance for contractor's profit but excluding preliminaries and VAT. Discussions on unit rate pricing led to a paper on the pricing of preliminaries, in which it was agreed which items would be included as preliminary items and which would be included with unit rates. The various discussions along the way have ironed out the difficulties and misunderstandings that have arisen.

An unambiguous list has been achieved for building work, but some difficulties still exist in regard to electrical and mechanical engineering items. Members were able to submit prices for these items in their own country's currency. This helped when comparing item for item in differing currencies but was little use for comparison of total project cost between countries. A method had to be found whereby the vagaries of currency conversion was avoided. This led to the development of **purchasing power relatives**.

10.4 PURCHASING POWER RELATIVES

Unit rates were applied to a quantity model for an office block based on a four-storey concrete-framed office block as described previously. The specification of the building is as follows:

- four-storey office, gross floor area 1,100 m²;
- reinforced concrete ground slab with edge beams;
- reinforced concrete stub columns on pad foundations with part reinforced concrete basement;
- reinforced concrete frame and floors;
- mansard-shaped steel roof covered with woodwool decking and asphalt;
- tank housing on roof of brick walls with felt on particle-board decking;
- brick and block cavity walls;
- aluminium double-glazed windows and entrance screen;
- brick and block internal walls, ply-faced flush internal doors;
- plaster and paint to walls, clay tiles on screed to ground floor, uncovered screed elsewhere;

Table 10.1 Purchasing power relatives 1991

Purchasing power 1991	United Kingdom	Rep. of Ireland	France	Netherlands	Denmark	Spain	Portugal	Belgium
United Kingdom	100.00	134.25	1512.50	420.00	1978.39	25,294.27	19,720.69	9885.54
Rep. of Ireland	74.49	100.00	1126.67	313.01	1473.70	18,841.69	14,689.92	7363.73
France	6.61	8.88	100.00	27.78	130.80	1,672.35	1,303.85	653.59
Netherlands	23.80	31.95	359.94	100.00	470.82	6,019.20	4,693.13	2352.56
Denmark	5.05	6.79	76.45	21.24	100.00	1,278.53	996.80	499.68
Spain	0.40	0.53	5.98	1.66	7.82	100.00	77.96	39.08
Portugal	0.51	0.68	7.67	2.13	10.03	128.26	100.00	50.13
Belgium	1.01	1.36	15.30	4.25	20.01	255.87	199.49	100.00

The table is read horizontally and is used to convert the cost of a building in the national currency of one country to the cost of the same building in the national currency of another country. For example:

a building costing £2M in the United Kingdom would cost 2,000,000 x 420.00/100 DFL in the Netherlands = 8,400,000 DFL

and

a building costing 2.5M FF in France would cost 2,500,000 x 8.88/100£IR in the Republic of Ireland = £IR 222,000.

- suspended ceilings throughout;
- rainwater, sanitary, hot and cold water, firefighting and electrical installations.

The item descriptions in the model are a compromise between the different design practices in each country, and each participating country has been asked to price the nearest common local equivalent to specifications given. The model assumes that offices are the same in each country. We know they are not. Nevertheless, the results give a reasonable indication of the relative cost of building in each country. The results of the priced model are used to calculate a 'purchasing power relative' (PPR) to compare costs between different countries, which is not susceptible to the vagaries of monetary exchange rates (Table 10.1).

Another advantage of the PPR is that as some countries have contributed this information over a five-year period it is possible to produce a time series index giving the annual building cost inflation figures for these countries as implied by the costs provided (Table 10.2).

10.5 EARLY COST ADVICE

It is quite apparent that the majority of countries use a number of different techniques to provide progressively more accurate information as the design of a building becomes more defined. It was thought that it would be helpful in understanding between countries if these techniques were noted.

10.5.1 United Kingdom

In the United Kingdom the building economist will be brought in at a very early stage, when the only information as to the needs of the client is some indication of function, size and quality or performance required of the proposed project. Drawings at this stage, if available, will be no more than simple dimensioned diagrams indicating the probable size and shape of the building.

The first and in some ways the simplest method of estimating is based on a **cost per functional unit**. Clients with a regular building programme of similar types of building (e.g. hospitals, schools or hotels) are usually able to set a budget or cost limit for a new project based either on their records of cost of similar buildings or on the rental income or sale value of the new project. Cost records will commonly be based on a cost per functional unit (e.g. £x per hospital bed, £y per school place, £z per hotel room). The

Table 10.2 Index series (base 1986 = 100)

	1986	1987	1988	1989	1990	1991
UK	100	102	113	118	108	97
Ireland	100	100	101	104	107	105
Netherlands	100	109	102	114	117	123
Spain	100	104	113	124	138	165
Denmark	100	106	113	116	124	127
France	100	104	107	111	115	116

budget or cost limit so calculated, owing to its speed and simplicity, is often used to provide a first approximate estimate or 'order of cost' estimate for clients who have not set a predetermined budget or who require a speedy cost of a project. Another method of arriving at this early approximate estimate is by cost per m² of gross floor area. If some idea of the floor area is known then cost per m² can be applied. These costs are available from a number of sources for a variety of different types of building.

The second method of providing early cost advice is based on the use of **elemental cost analyses** of earlier similar projects, making appropriate adjustments to produce a cost plan of the proposed new building. The adjustments to the cost analyses selected would include: quantitative adjustments to allow for different floor areas, wall finishes, services etc.; qualitative differences to allow for different wall construction, roof finishes etc.; and price levels to allow for general price inflation and location differences between the projects.

The third method, **approximate quantities**, is only practicable if some sketch plans or diagrams are available. The technique consists in measuring the quantities of work in the new building and pricing them based on previous projects suitably adjusted for inflation and location etc. A new method has been launched by the RICS. It is an 'expert system', which uses intelligent knowledge-based system (IKBS) technology to produce a reliable framework and working programme for strategic planning of construction proposals.

10.5.2 Holland

In Holland re-use of project information is not common at a global cost level because, it is claimed, of the unique status of each project. Information is available at elemental level, and this is the most common method of producing estimates. In 70% of projects early cost advice is available at bills of quantities level; for the remaining 30% it is available at elemental level.

10.5.3 Spain

In Spain early cost advice is available at inception and at 'avant project' phase. At inception the cost is determined by comparison with historic costs of similar buildings. At the 'avant project' phase concepts of form, geometry and quality of functional elements are all required. With these defined, an elemental costing can be carried out, which investigates the relationship between the functional elements and gross floor area. The cost of different design solutions can also be evaluated.

A system has been developed that uses a cost model to analyse cost and to evaluate different design solutions. Coefficients are given to variables in the **images of reference** that have been established for different building types. The system develops the model in stages and can reconcile different solutions to the main aim of the building.

10.5.4 Finland

In Finland approximate estimates require drawings, while the cost plan would normally be produced before drawings were available. The main method used to arrive at a budget is based on cost per m² for different categories of rooms or buildings. Previously, cost of functional units were used before the present system was developed.

10.5.5 Switzerland

In Switzerland, traditionally, early estimates for building work are based on the volume of the building and the cost per m³ of comparable finished buildings. Lump sum or percentage additions are made for site costs, preparatory works, special equipment, furnishings, external works and incidental costs. These estimates are sometimes subdivided on a percentage basis to provide target costs for trade sections. A detailed estimate will normally be produced towards the end of the design stage (shortly before production of final working drawings) based on simplified bills of quantities. Increasingly, elemental estimating techniques are being used based on the Swiss standard form of **cost analysis** (BCA). These estimates are based on historical data from comparable projects, with the necessary adjustments for quality and price.

Another system is also used, based on **elemental approximate quantities**. This method, which lies somewhere between elemental cost planning and approximate quantities, is a computerized system that synthesizes rates based on trades for each sub-element. The information can then be re-sorted into bills of quantities items or functional elements.

10.5.6 Germany

In Germany, three stages are in common use for early cost estimates:

- In most cases, for simple buildings or private clients, a system is used based on cost per m³ or m² using prices for similar buildings in the recent past. Cost estimates at this stage can also be made on a functional unit cost for buildings such as old people's homes, hospitals, hotels etc.
- Estimates for rough building elements are based on sketch plans, and provide a cost framework for the building. The estimates can be used for rental considerations by private clients or for further planning. Adjustments for quality and price can be made to the cost plan as necessary.
- Estimates for detailed building elements divide the building into elements with quantities measured from drawings. Unit rates for each element are taken from a database and adjusted as necessary. From these cost estimates the bills of quantities can be produced on a trade basis by re-sorting of the items from the cost estimate.

10.5.7 Denmark

In Denmark early cost advice is based on an **uncertainty curve**. In the early stages of a building the uncertainty is high; as design progresses the uncertainty becomes less. The SfB classification is used for elemental cost advice, with range estimating using minima and maxima to give a mean price together with minimum and maximum prices for a project. The use of the standard deviation allows the calculation of the probability of any one project being close to the mean figure. As the design progresses, estimates proceed using a space/room programme, functional diagrams, life-cycle costing and time planning.

10.5.8 France

In France **elemental cost planning**, is used with the costs broken down into some 30 functions. The system, which is computerized, uses synthesized elemental cost to produce a cost plan. The cost plan is then used for cost control purposes.

10.6 FUNCTIONAL UNIT COSTING

The Cost Commission thought that as a number of countries used **functional units** (FUs) in some form or other, it would be beneficial if agreement could be reached on a European defined list of FUs for various building types. So a start was made to this end, although progress to date has been slow, owing to concentration on other studies. The concept of FUs and FU costing is not new, and has been used for a considerable number of years. In fact it was in existence when Christopher Wren built St Paul's Cathedral.

A number of government departments in the UK produce FU costs: for example, the Ministry of Agriculture, Fisheries and Food produces a guidebook on the cost of a number of specialized farm buildings. The Home Office has developed a system of FUs for costing prisons and police buildings. The Department of Health has a comprehensive system of costing FUs for hospital departments and health buildings. The Department for Education issues cost guidelines that relate to **area-based cost units** (ABCUs) for different types of schools, higher education establishments and university buildings. The Department of the Environment produces indicative total costs for housing, expressed on a per person dwelling basis. Various other types of building all have some form of FU costing.

FUs have a number of uses:

- for estimating the cost of a project at an early stage based on minimal information;
- for setting a notional cost plan;
- for setting a project budget cost;
- as a quick comparison between similar projects.

It is hoped to continue with this study if resources permit.

11

Global costs (life-cycle costs)

Jacques Moreau

11.1 INTRODUCTION

This chapter is not a guide for calculating costs. It serves only as a collection of definitions and as a model framework for the component parts. Produced in cooperation with all the members of the European Committee of Construction Economists, its objective is to permit each country to establish a common method of calculation, adapted to the customs and uses of each country, but allowing each to obtain coherent and compatible results with a new European harmonization. The Global Cost Commission hopes to meet in a few years to compare the detailed methods originating from this matrix and the results achieved.

11.2 DEFINITIONS

11.2.1 Global cost

The building global cost includes:

- the initial investment cost;
- annual running cost:
  - maintenance expenditure,
  - working expenditure;
- the resale cost.

11.2.2 Estimated global cost

This is the sum of the approximate costs of construction and operation of a building during a specific period. The sum is calculated at constant prices.

11.2.3 Economical global cost

Actual global cost is the estimated and global cost set out listing actual expenditure and including all or part of such accounts as:

- taxes and income taxes;
- loan interest;
- price fluctuations.

11.2.4 Reference period of a global cost calculation

This is the number of years of maintenance and operation taken into account in the calculation. This period depends on the type of investor (seller, lessor or owner). The calculation may be based on periods varying from 10 to 50 years.

11.2.5 Duration of life

This is the period before replacement is anticipated. In most cases, this period is a function of the maintenance quality. It is also necessary to take account of the possibility of repairs during use.

11.3 UTILIZATION (USE)

First of all, the calculation of a global cost is a decision-making aid, which allows the client to:

- appreciate the profitability of his investment;
- choose from several technical solutions the one that is the most economical;
- choose the best solution for financing.

A calculation of global cost should be made as part of an initial investment decision. The amount of the global cost should be updated during the project period.

11.4 QUALITIES OF A CALCULATION SYSTEM

A calculation system should have the following qualities:

- speed of data acquisition;
- freedom to include or omit data;
- possibility of variations of times, rates, costs for a quick simulation of results given various assumptions;
- reliability of results

11.5 RELIABILITY

The operating costs and especially the maintenance costs can be only relied on if the conditions of use and the maintenance are respected by the users.

11.6 PRESENTATION

11.6.1 Estimated global cost

The costs are staggered by years; 0 is the moment of the end of the initial investment, and 1 is the first operating year:

Cost	Years									
	-3	-2	-1	0	1	2	3	4	5	etc

Note: The initial investment may take more than one year.

11.6.2 Economic global cost

These expenses are staggered by actual dates of payment. Year 1 is always the first year of operation. The initial investment will be distributed between purchase date of the land and the loan refunding date, if necessary. The table can be completed by an annual re-allocation of periodic receipts, which enables the calculation of the investment rentability.

Expenses	Years									
	−3	−2	−1	0	1	2	3	4	5	etc
		Investment					Return			

11.6.3 Taxes

It can be useful, for particular clients, to identify for each expense the VAT amount at the time of payment. The repartition of the receipts will identify this VAT at the time of its recovery.

11.7 COMPONENTS OF A TECHNICAL GLOBAL COST

Components	Comments
A INITIAL INVESTMENT	
1 **Preliminary studies**	
2 **Site acquisition** including the cost of the building in the case of refurbishing	Initial investment without effects on the construction or operating or maintenance costs
3 **Preliminary and associated expenses**	
3.1 Preliminary expenses • Land survey – soil exploration analysis, findings • Advertising – duplication • Demolition – compaction • Temporary installations – fences	

Components	Comments
3.2 Taxes and indemnities • Acquisition of party walls • Connection taxes (telephone) • Taxes to developers organizations	For some countries only
3.3 Services Connection work cost outside the property limit	
3.4 Commissioning costs Testing and checking before putting into service	For some countries, or some projects only

4 Construction itself

Components	Comments
4.1 Construction itself 411 Substructure 412 Superstructure Frames Roofing External walls Stairs and balustrades 413 Equipments Structural equipments Services equipments Finishing	Partitioning of rooms, coverings of floors and walls and roofs
4.2 Additional costs for adaptation of the site 421 Site preparation 422 Specific foundations 423 Service system • incoming • outgoing 424 External works • Access • Fences Type Walland • landscaping • outbuilding	

Components	Comments

4.3 *Specific equipments*

431 Specific materials and systems
432 Specific protection
433 Specific equipments for building
  purpose
434 Specific furniture
435 Pool and other aquatic
  equipment

5 – Indemnities

5.1 *Taxes*   In France: taxes for increasing the 'COS': coefficient of ground occupancy taxes for creation of offices in housing location

5.2 *Eventual subsidies*   For expenses deduction

6 – Fees, insurances

6.1 *Professional fees*

6.2 *Technical control fees*

6.3 *Testing costs*   Public or private organizations, according to the country

6.4 *Investor insurance*   In France: insurance for building damage

6.5 – *Various advising fees*
The initial investment is defined as
the sum of items 1–5 above.

B – AGREED COSTS

7   Exploitation expenses

7.1 *Building and specific equipment*   Possibility of two different points:
711 Superstructure window cleaning   7A owner's expenses
712 Structural equipments floors and   7B tenant's expenses
  walls cleaning

Components	Comments
713 Functional equipments	For all of these expenses, add if necessary the maintenance contracts for the equipments (cleaning system, etc.)
• cold water: consumption treatment	
• waste water: treatment	For industrial or office building, exclude the costs of energy for the production equipment and the telephone charges, which are expenses independent of the building life
• heating: energy maintenance	
• ventilation: energy maintenance	
• electricity: consumption relamping	
• generator	
• low voltage installations	
• lift apparatus	
• other technical equipment	
7.2 Outside Landscaping maintenance	
7.3 Caretaking Wages and social expenses Maintenance of the premises	
7.4 Management Wages and social expenses 741 Intelligent building management system 742 Building management	
7.5 Special taxes for development	Including administration For some countries only
8 Maintenance	
8.1 Preventive maintenance Constant annual lump sum	Incidental expenses that, if they are not respected, could have important effects for points 7.3 and 7.4
8.2 Current maintenance Constant annual lump sum Repairing exterior painting	Replacing the glazing, repairing the toilets, the ironmongery etc. Replacement frequency to be estimated

Components	Comments
8.3 Major repairs, replacement and cleaning	
831 – on external building envelope walls and roof covering random equipment on equipment on external works road restoring planting – landscaping	Replacement of parts of equipment (motors, cables, taps etc.)
832 Total replacement • Roof • External woodwork • External coatings • Finishing of floor, walls, roof • Internal joinery • Technical equipment • external works	
9 – **Resale value**	Chargeable for the previous year; generally limited to the value of the land

11.8 COMPONENTS OF AN ECONOMICAL GLOBAL COST

Components	Comments
A – EXPENSES	
5 – *Initial investment* (Part 1 to 5) Distributed among • Equity capital • Borrowed money with interest	Chargeable to the actual year of the expense
6 – *Exploitation expenses* Part 6 annual amount (with inflation)	Possibility of simulation with different rates of inflation
7 – *Maintenance* Annual amount with inflation	

Components	Comments
B – INCOMES	
Sale	Different assumptions of repartition in
or	time
Annual rents (with a calculation of inflation) and a factor	
or	
Incomes or tickets for	
Resale	Initial value with assumption for inflation
A – B = *CASH FLOWS*	

12

Facilities management

David Owen

What is facilities management? We shall try to find an answer to this question by considering the following issues:

- definition;
- history;
- concept;
- scope;
- examples.

12.1 DEFINITION

There are many different definitions of facilities management. For example:

- the practice of coordinating people and the work of an organization into the physical workplace (IFMA);
- the practice of coordinating the physical workplace with the people and the work of an organization, integrating the principles of business administration, architecture, and behavioural and engineering sciences (Library of Congress);
- the development, coordination and control of the non-core specialist services necessary for an organization to successfully achieve its principal objectives (Symonds FM);
- the process by which an organization delivers and sustains agreed support services levels within a quality environment to provide full value in use to meet strategic objectives (Strathclyde CFM).

However, simply referring to one of the many definitions is not the best way to understand facilities management. Why? As we shall see, facilities management is a relatively new concept – one that is evolving rapidly – and there is some justification in the criticism that it is all things to all people. Hence there are a great variety of definitions, many of which attempt to cover too much ground, and end up being vague.

Consequently, the understanding of the term 'facilities management' tends to vary according to the background of the source asked. To someone from telecommunications, facilities management is concerned with the management of the telecommunications network of an organization; to a services engineer, it relates to the maintenance of the building's mechanical and electrical services.

From this diverse background, you will begin to understand why a commonly agreed definition is difficult to achieve. Computer scientists, building designers, catering contractors, etc. have little in the way of a common vocabulary. However, at this early stage, let us be clear that facilities management is not *just* building maintenance management or service maintenance.

Another reason for confusion is that property and construction-related people dominate the supply side of the industry, whereas the demand side of the industry is staffed by those with generally poor awareness of property-related matters: for example, office administrators and general managers.

I would argue that gaining an understanding of the concept of facilities management, its history and its scope will give a much clearer insight and may allow us to put our own definition forward for consideration later.

12.2 HISTORY

I believe it helps our understanding of facilities management if we are aware of its roots and its development to date. There is a remarkable amount of consensus regarding the origins of facilities management. This is only fully appreciated when both the diversity and range of its constituent parts and the speed of its growth are grasped.

Facility management, as a description, originated in the information technology industry. The term was coined in 1964 in the USA by the founder of the concept, Ross Perot, who established EDS, an organization that now employs approximately 64,000 staff worldwide in a US$6 billion p.a. business. The same term was then borrowed from the information technology industry and given much wider prominence by the forming of the Facilities Management Institute (FMI) in 1979. FMI was an offshoot of parent company, Herman Miller. This later developed into an independent organization known as the International Facility Management Association (IFMA), which is now based in Houston, Texas. In the USA there are currently about 11,000 qualified members of IFMA, with an estimated 56,800 using the job description of Facilities Manager.

In Europe – in the UK for example – the position evolved less rapidly. One reason for such confusion as exists lies in the fact that there are two organizations vying for pre-eminence:

- The Association of Facilities Management (AFM);
- The Institute of Facilities Management (IFM).

The AFM is an independent institute, the IFM is a subculture of the Institute of Administrative Management (IAM). Nevertheless, major strides in the development of facilities management have been made in the UK, particularly in the last two to three years, owing largely to an informal system of networking between the principal protagonists.

Euro FM was formed in 1990 as a network of parties interested in encouraging the spread of the understanding of facilities management throughout Europe.

12.3 CONCEPT

There are only two words to analyse:

- facilities (plural of facility);
- management.

In reverse order, **management** comprises the functions of:

- planning;
- organizing;
- staffing;
- directing;
- controlling.

According to various dictionaries, **facility** is the fact or condition of being easy or easily performed; freedom from difficulty; opportunity for the easy or easier performance of anything; easiness to be performed and dexterity; readiness of compliance. None of this helps us at all, but it does illustrate just how new this terminology is – it's certainly ahead of the dictionaries.

'Facilities' usually means any property where people are accommodated, usually where they work, or where an organization conducts its business, which includes office, manufacturing, R&D, residential, health, educational and leisure facilities (AFM).

Put together, **facilities management** is a management tool or system which can be employed at all three levels of an organization:

- strategic;
- tactical;
- operational.

The activities and the physical facilities will have all existed before facilities management was considered. The primary concept is the coordination of people and the workplace to support the business – this is what is new.

12.4 SCOPE

Partly because of facilities management's evolving definition (as opposed to rigid definition) and partly because of the fact that it started by identifying a real business need, facilities management has attracted many market sectors to it: some looking for identity, some attracted by success.

At its simplest level, facilities management encompasses:

- premises (buildings);
- support services;
- information services.

But you can assume that every organization will take a differing view of what constitutes facilities management. My position is to accept this fluid state and embrace it with an understanding of the holistic view.

For example, those of us who come from the construction industry must make an effort to see this concept from the client's viewpoint. If your interface with a client organization (which I will now call the 'user') is the facilities management function, then you are just as liable to be communicating with someone from an administration background as from a catering or security background. They are unlikely to understand property matters and are less likely, again, to be conversant with construction procedures. However, the opportunity exists, if you talk to them in their own non-property language, to demonstrate the added-value that you can bring to a range of construction and property issues. The danger is that you will speak to them in your 'alien' construction language, showing that you don't understand their business needs or problems.

It is essential that this opportunity is approached in a market-led, demand-led manner; not supply-led. In a recent article in the *Chartered Quantity Surveyor*, facilities management was described as 'the control and most appropriate use of property resources': in itself not a bad proposition, but the article immediately went on to list the following aspects as 'encapsulating facilities management':

*   space planning;
*   space costing;
*   asset tracking;
*   life cycle costing;
*   maintenance;
*   component specifications.

All these aspects certainly do fall into the orbit of facilities management. However, by themselves, they describe an ultra-narrow view of the scope of facilities management and hence the market, and, most importantly, the scope of opportunities for suppliers.

To comprehend the full scope, it is interesting to consider whether there is any difference between the way in which facilities management is developing in Europe compared with the USA. I consider there is a different emphasis emerging here in Europe – one that I believe makes facilities management an even more exciting proposition.

IFMA lists the following nine functional responsibilities as the principal constituents of facilities management:

*   long-range planning;
*   annual planning;
*   financial forecasting and budgeting;
*   real estate acquisition/disposal;
*   interior space planning;

- architectural, engineering, planning and design;
- new construction and/or renovation;
- maintenance and operations management;
- telecommunications integration, security and general administrative services.

They go on to complicate it a trifle by subdividing into 1600 subcategories.

You will see a distinct emphasis on space utilization and planning and, hence, on buildings: perhaps an ideal bandwagon for construction economists to climb on. But I think it gets better in Europe because of the shift towards an even broader scope. This involves an acceptance of the core and non-core business philosophy. Europe may have lagged behind the USA initially, but there is a growing awareness that facilities can be managed to add value and add quality to a user's organization.

The driving factors bringing about these positive changes include such macro-economic pressures as:

- global competition;
- global change;
- information technology advances;
- more sophisticated user expectations.

In micro-economic terms, these macro-influences have caused organizations to focus on core business as never before, in order to improve competitiveness in quality and value terms.

I think it is a pity that the structure of an organization is divided between 'core' and 'non-core' business. 'Core' business is a tremendously expressive term but 'non-core' sounds negative and, perhaps, unnecessary. While nothing could be further from the truth, I find it easier to think of an organization in terms of 'core' and 'essential support services'.

What is core business? Core business is the primary function/functions or process/processes of an organization: that is, it is the reason for its existence. Consequently, non-core business becomes everything else that is necessary to support the reason for existence.

12.5 EXAMPLE

I see the facilities management function of an organization as being the interface with a knowledgeable client, possessing a wide range of wants.

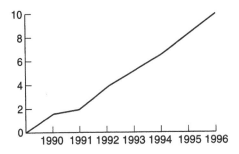

Fig. 12.1 Value of IT facilities management in Europe in US$ billions.

Understanding this will help to unlock a broader range of opportunities for suppliers of technical and professional services than previously thought possible. To give one example, I shall briefly introduce you to one business strategy that facilities management makes possible.

Contracting-out is the process by which a user employs the services of an out-of-house contractor. This is often known as 'outsourcing', although strictly outsourcing requires a transfer of people and other assets (such as machinery) to the contractor.

In the limited space available, let us look at one example of the European experience – taking information technology (IT) as the object of the con-tracting-out. The Computer Sciences Corporation Index Survey of European Information Systems Executives found that 71% are *planning* to contract-out some IT operations, compared with 36% in 1990/91. This will boost the value to the contracting-out market from US$1.6 billion in 1990 to $10 billion by 1996.

Figure 12.1 shows the growth achieved and the potential in this particular sector of facilities management. The pie chart in Fig. 12.2 illustrates the way in which contracting-out has been accommodated in the IT sector in Europe.

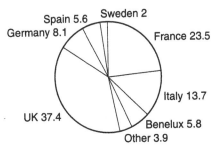

Fig. 12.2 Share of IT facilities management contracts by country (%).

I have deliberately used IT (or IS) as the example for two reasons:

- It should consolidate the view that facilities management is *not* just about buildings and construction.
- Study of the history of facilities management shows that the other sectors of premises and support services follow the direction taken by the cutting edge made by IT.

12.6 SUMMARY

If nothing else, the world recession has encouraged, if not forced, organizations to reconsider their core businesses, and with this has grown the importance of coordinating the non-core. This has resulted in the raised importance of facilities management as a strategy to improve an organization's performance and value.

Don't worry too much about a definition for facilities management – recognize what it *means* and its *scope*. The scope of facilities management is extremely broad. Because it is an evolving and developing area there is a tendency for it to be all things to all people. However, once the scope is grasped, it will be seen that, because of the dominance of property in the facilities management market, construction economists are ideally placed to capitalize on the opportunities afforded.

Facilities management means that companies and public authorities (such as the health sector, education sector, and government departments) are rapidly coordinating their own in-house non-core business structures, presenting an in-house one-stop shop, and thus giving consultants and contractors the ideal target interface for a wide range of procurement possibilities. Further, with continued pressure to concentrate on core business, the awareness of facilities management is variously forcing or encouraging organizations to 'down-size', with contracting-out being an optimum method of reducing non-core staff and overheads, to the benefit of both the organizations (the users) and the supply industries.

Finally, that attempt at a definition:

Facilities management is the active management and coordination of an organization's non-core business services together with the associated human resources and its buildings, including their systems, plant, IT equipment, fittings and furniture, to assist that organization to achieve its strategic objectives.

13

Cost planning and controlling building costs

Ari Pennanen

13.1 PURPOSE OF COST PLANNING

The purpose of project management is to achieve all the targets set for the project with a reasonable quantity of resources, and it should enable the owner to undertake long-term budgeting. Targets are usually the rooms needed, acceptable costs and time schedule.

Cost planning is operative work done by the project manager and the project organization to reach the financial targets.

13.2 VARIATION OF BUILDING COSTS

To manage our difficult task, cost planning, we have first to look at what causes the large variation of building costs in different buildings.

13.2.1 The origin of the costs

Building costs are, in general, caused by the amount of resources that are fixed to the building during the process and by their unit prices. The resources are work, materials and energy. This is how the costs come true, and to find out the costs in different buildings we have just to estimate the drawings.

From the cost planner's point of view, estimating the drawing is, how-ever, a passive way to operate with the costs, even though it is important.

By just estimating drawings you are always too late. You have to find out what actually influences the use of resources.

13.2.2 Function of the rooms: the owner's point of view

The owner's point of view is rooms, when a building is considered. The owner sees a building as the rooms in which he can live or do business. He requires different features for different needs. Often it must be warm in the rooms; sometimes rooms must be high; sometimes he requires the possibility of washing hands. The owner often sets requirements on the atmosphere of the room and its environment.

The owner is not actually interested in whether the partition wall is made of plasterboard or other materials, as long as it meets the requirements set: on sound insulation, for example. The column may be made of concrete or steel; the owner doesn't mind.

The owner's point of view fixes the quantities and unit prices of the building elements. If the owner wants more height in the room, the builder builds more partitions, columns, external walls and finishing. If the owner requires sound insulation, the builder makes the partitions heavier. If the owner wants to wash his hands in the room, the builder connects water pipes to the tap and basin (Fig. 13.1).

Rooms, and the requirements on them, are the main cause of the variation

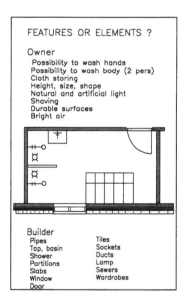

Fig. 13.1 Features or elements?

of costs. Because of the variation of requirements a building might cost £150/m² (e.g. unheated storage) or £3500/m² (e.g. a laboratory).

13.2.3 Conditions

The conditions on the building site may create requirements for resources that are not related to the requirements of the rooms: piling, stabilizing or quarrying, for example.

13.2.4 Design solutions

Given the same rooms, requirements and conditions, different designers will do the principal layouts and final details in different ways, and the result is different quantities and different unit prices. The variation of the costs caused by different design solutions might be quite large: ±15%, for example.

13.2.5 Refurbishing

The results of refurbishing, like those of new buildings, are rooms. The owner's point of view determines the required quantity of resources when the result is considered. When rehabilitating, we spend resources if the future function of the rooms requires features that cannot be found in the existing building. This may be caused by a change in the function of the room. It also may be caused because the elements that provide the features are broken or in bad condition.

13.2.6 Other factors

Other factors that cause variation in costs include:

- local variation of costs, mainly due to variation of labour costs;
- inflation;
- market situation.

13.3 COST PLANNING

13.3.1 Budgeting

In cost planning we need a target and a programme of work to achieve that target. The target is the budget, which must be fixed before the design has started. The rooms, and the requirements set on them (the owner's point of view) are the main cause of the variation of costs. So the owner's point of view is most relevant at the budgeting stage.

The budget must be based on the same information that is given to the architect and other designers when planning starts: the owner's requirements. It should not get any information from the drawings, otherwise the target for designing would be the designs themselves. Of course all the other factors known before designing starts must also be taken account: site conditions, local factors, inflation, market situation, and existing building (in refurbishing work).

Budgeting is not estimating with inadequate information. We have all the information that we need.

13.3.2 Keeping to budget

Design solutions causes variation in building costs. But because we don't get any information from the designs when budgeting, the design solutions do not influence our budget. Instead, you have to influence the plans. Steering the design can be done all the time during the design stage, but the easiest way is to do it in the very early stage of designing.

13.3.3 Responsibility of cost planning

Cost planning in a building project is a question of attitude. It is possible to keep to the budget if you determined to achieve the targets.

Steering the design must be done by all the project organization (project manager, architect, engineers etc.). It is the project manager's duty to motivate the project organization to the common targets and to work for the targets.

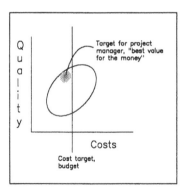

Fig. 13.2 Quality and building costs.

13.3.4 Quality and costs

In building design it is possible to achieve good quality if the designers are good, if you give them a chance to work creatively, and if the project organization is motivated. Creative work is made possible if the requirements can be reached by several alternative solutions. So the budget should not be too low. The budget should not be too high either, because this obviously leads to unreasonable use of resources, and does not guarantee quality.

The correlation between quality and building costs is quite weak (Fig. 13.2). If the budget is set at an average level in regard to the requirements, it is possible to find numerous different design solutions, and almost whole

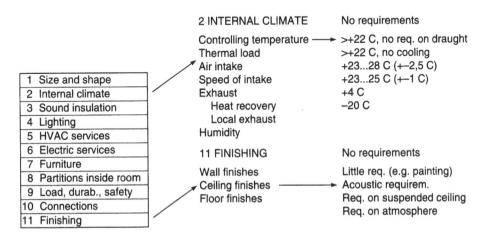

Fig. 13.3 Choosing the requirements of a room when budgeting. The target price method transforms the requirements into money.

the range of possible quality can be reached. You just need work and skilled people to find a solution with quality and reasonable costs.

13.4 TOOLS NEEDED IN COST PLANNING

In cost planning, the most important element is attitude. However, you need also tools that help you in budgeting and in steering the design.

13.4.1 Budgeting

A budget, done properly, is not a forecast. It is a target, which can be reached by methods of cost planning without losing the other requirements. Because the needs and requirements of the owner cause the quantities and unit costs of the elements, the owner's point of view is useful when budgeting. It is possible to find out the needs of owner – the rooms, sizes and features – at a very early stage. In refurbishing also, the result is rooms. The owner's point of view determines the needed quantity of resources for the result. To find out how much resources we need when refurbishing it is useful to use the builder's point of view again. If a room does not provide the features needed you have to produce them. After knowing the requirements of the owner they must be transformed into money. In Finland the **target price method** provides the project manager with advice on changing the owner's requirements to produce the budget (Fig. 13.3).

The target price method is based on the rooms and on the requirements set on the rooms. The method is used to create standards of quantities and unit prices of the resources with which you can produce different features. For instance the required size, shape and height fix the quantity of the partitions. The required standard of sound insulation fixes the unit price of the partitions. If you need to wash your hands in your room, the method provides you with a standard of pipes, basin and tap etc. The method uses only the language of the owner. You don't need to know anything of the building elements to produce the budget.

The target price method can be used when building new buildings and also in rehabilitation projects. Budgets can be produced manually, but usually target price software is used. The target price method provides a budget set at an average level in regard to the requirements. About half the sketches fit with a budget and half exceed it. The project manager has to make sure that his project does not exceed budget at the end of the project.

TECHNICAL COLLEGE IN HELSINKI, ROOM SCHEDULE

Room schedule	5 238	Rooms altog.	7 178	Unit price	60
Traffic spaces	1 616	Structures	951		
Technical rooms	324	Grm2, target	8 129	45 660 694	106

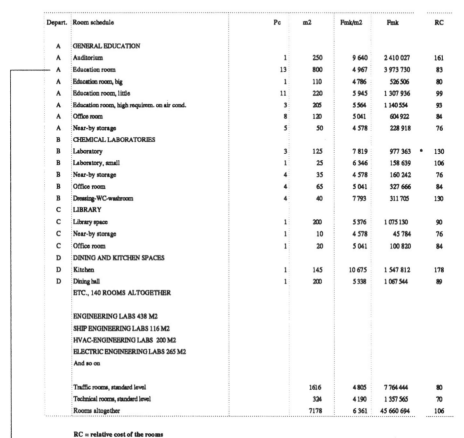

Depart.	Room schedule	Pc	m2	Fmk/m2	Fmk	RC
A	GENERAL EDUCATION					
A	Auditorium	1	250	9 640	2 410 027	161
A	Education room	13	800	4 967	3 973 730	83
A	Education room, big	1	110	4 786	526 506	80
A	Education room, little	11	220	5 945	1 307 936	99
A	Education room, high requirem. on air cond.	3	205	5 564	1 140 554	93
A	Office room	8	120	5 041	604 922	84
A	Near-by storage	5	50	4 578	228 918	76
B	CHEMICAL LABORATORIES					
B	Laboratory	3	125	7 819	977 363 *	130
B	Laboratory, small	1	25	6 346	158 639	106
B	Near-by storage	4	35	4 578	160 242	76
B	Office room	4	65	5 041	327 666	84
B	Dressing-WC-washroom	4	40	7 793	311 705	130
C	LIBRARY					
C	Library space	1	200	5 376	1 075 130	90
C	Near-by storage	1	10	4 578	45 784	76
C	Office room	1	20	5 041	100 820	84
D	DINING AND KITCHEN SPACES					
D	Kitchen	1	145	10 675	1 547 812	178
D	Dining hall	1	200	5 338	1 067 544	89
	ETC., 140 ROOMS ALTOGETHER					
	ENGINEERING LABS 438 M2					
	SHIP ENGINEERING LABS 116 M2					
	HVAC-ENGINEERING LABS 200 M2					
	ELECTRIC ENGINEERING LABS 265 M2					
	And so on					
	Traffic rooms, standard level		1616	4 805	7 764 444	80
	Technical rooms, standard level		324	4 190	1 357 565	70
	Rooms altogether		7178	6 361	45 660 694	106

RC = relative cost of the rooms

Unit price = unit price of relative cost

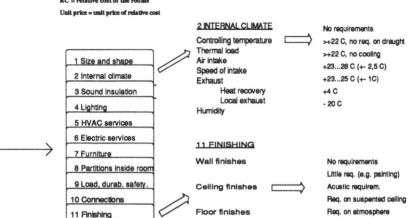

1 Size and shape	**2 INTERNAL CLIMATE**	
2 Internal climate		No requirements
3 Sound insulation	Controlling temperature	>+22 C, no req. on draught
4 Lighting	Thermal load	>+22 C, no cooling
5 HVAC services	Air intake	+23...28 C (+- 2,5 C)
6 Electric services	Speed of intake	
7 Furniture	Exhaust	+23...25 C (+- 1C)
8 Partitions inside room	Heat recovery	+4 C
9 Load, durab. safety.	Local exhaust	- 20 C
10 Connections	Humidity	
11 Finishing		

11 FINISHING

Wall finishes → No requirements / Little req. (e.g. painting)

Ceiling finishes → Acustic requirem. / Req. on suspended ceiling

Floor finishes → Req. on atmosphere

Fig. 13.4 Budgeting a technical college in Helsinki.

Figure 13.4 shows part of the process of budgeting for a technical college in Helsinki. The budget is based on a room schedule for the college. The required rooms are first chosen from the list of rooms in the target price method (nowadays there are about 500 room standards included in the method). Part of the room schedule is shown. You can pick up any room and change any requirements set on the room. For example, if you required very strict temperature controlling in the education rooms, the unit price of education rooms (Fmk/m²) would rise (nowadays there are 80 choices concerning the requirements in the method).

13.4.2 Estimating the sketches

The owner's requirements can be met by an infinite number of architect's design solutions. Architects do not usually know whether their first plans will lead to an expensive design or not. You should not change your budget because the first plan happens to be expensive. Instead, you can influence the design process, especially in the early stage of design.

The project manager has to help the architect to find the design solution that reaches both the owner's requirements and the budget. He has to be sure that he, the architect and the owner have a common target. He has to give immediate feedback about the expense of the plan at a very early stage of the design.

The project manager does not let the design proceed from the sketches until the architect has found the plan that fits with the requirements. The

		Unit	Quantity	fmk/unit	fmk		
37	Roofing structures	m²	1425		206 625	1 417	143
	Wooden trusses	m²	1425	145	206 625		
372	Eaves	jm	312		54 564	318	112
	Wooden eaves 0.6 m, no gutters	jm	102	82	8 364		
	Wooden eaves 0.6 m, gutters	jm	210	220	46 200		
51	Roof deck	m²	1425		142 500	1 417	84
	Clay tiles	m²	1425	100	142 500		
4	Supplementary structures				593 280		
41	Windows	m²	450		593 280	441	1 255
	Wooden frame 131.3 gl., av. size 0.5 m²	m²	352	1 240	436 480		
	Aluminium frame, av. size 1.0 m²	m²	98	1 600	158 800		

Fig. 13.5 Estimating the sketches. Part of an elemental estimate. The estimate is produced by measuring the elements and finding their unit prices from the register. In the grey area the cost planner sees the quantities and unit prices that the target price method has created from the room schedule and requirements set on the rooms. This may help estimating in the very early stage when you cannot measure all the elements.

early stage of design is a creative stage of cost planning. To give economic feedback a project manager has to have the ability to estimate the sketches. He needs suitable tools again. In Finland, sketches are estimated by a method based on building elements (Fig. 13.5).

14

A model of professional training for a European construction economist

Antonio Ramirez

14.1 INTRODUCTION

This work is included within the grand objectives stated in Article 1 of CEEC Statutes which, as stated, are 'to promote training and qualifications of persons who are responsible for construction economics and to draw up proposals for the harmonization and acceptance of standards of training and qualification'.

With the intention of fulfilling such an important aim, the CEEC General Meeting, held in Copenhagen on September 1988, entrusted the Spanish delegation with the task of proposing a study programme, 'to explore the possibilities of harmonization of methods, training programmes and procedures among the member bodies'.

Our work is a synthesis of the document presented by the Spanish delegation to the CEEC meeting in Lisbon, April 1989, which was approved in Paris, October 1989. This document, entitled *Proposal for a study programme for a European Construction Economist*, included the observations made by the French delegation at the CEEC meeting in Seville and the information included in the document entitled *The core of skills and knowledge base of the quantity surveyor*, elaborated by the RICS and presented at the Qualifications Commission held in Paris on 16 October 1992.

Several important matters are developed in this document. The core of skills comprises:

- **Knowledge base.** Construction technology, measurement rules and conventions, construction economics, financial management, business administration and construction law.
- **Skills base.** Management, measurement, analysis, synthesis and communication.
- **Markets**. By services provided, such as value management and procurement management; by market sector, such as construction engineering and property; by geography.
- **Constraints**. The diversification of professional services to meet market demands is constrained by a number of factors, including the traditional boundaries in which a professional operates.

For example, the limits of professional knowledge and market information are meant to delineate a point beyond which professionals would cease to offer services as competent individuals.

Based on these elements, three objectives are determined, as follows.

- Identify key questions for research into knowledge and skills that the profession requires to develop its services.
- Identify the role of market needs.
- Identify the role of constraints.

Finally, in the chapter of conclusions, there is an important reference to the implications for professional education, where it states:

> The pressure of technological change means that there will be a demand for specialists. On the one hand, the increasing technological complexity of construction means that there is a need for specialization. On the other, there is a need for generalists to manage the interfaces between specialists on complex projects. These skills may or may not reside in the same individual. It is clear that professional education will need to encourage entrants from a range of professional backgrounds, perhaps taking specialized training in quantity surveying on top of a non-cognate degree.
>
> There is a need for an increasing emphasis on management and business skills in the technological context of construction. In the short term there will remain a need for graduates to be familiar with certain transient skills and techniques used in quantity surveying practice. The nature of these skills will change quite rapidly over time as information

technology continues to penetrate the profession. Professional education must enhance its focus on management, analysis, appraisal and communication (undoubtedly necessary for survival) in the context of the technology of the industry.

One important problem is to maintain an appropriate balance between certain fundamental, non-transient areas of knowledge and the technological context in which this knowledge will need to be applied. In particular, we are referring to the need to avoid allowing the, as it were, tail of technology from wagging the dog of knowledge. Many people outside construction, both at home and abroad, regard undergraduate courses in surveying as too narrow and vocational and too closely linked to long-established roles and professional institutions.

The reference framework that has been stated provides a variety of points of view that are difficult to harmonize into a common body, enriched with the singularities of each model. The analysis of all contributions allowed the formalization of an inventory of activities and functions of the construction economist, which are listed in the proposal for a professional profile of the construction economist.

From the functions listed in the profile, we have developed a general plan for training, divided into six areas of theory, which comprise all the fundamental matters, and a complementary area of practical training. In the document contributed by the RICS, a reference to the specific study of communications is included in the basic training to facilitate relations with the plurality of markets within the European Community, starting 1 January 1993.

The next step was a global proposal assigning the amount of time appropriate to each subject.

Finally, to facilitate interpretation, we have synthesized the planning and time and presented a summary in a chart (Fig. 14.1).

THEORETICAL TRAINING AREAS

	1. Basic	t1	2. Technological	t2	3. Legal	t3	4. Mathematical	t4	5. Accounting	t5	6. Construction Economics	t6
	1.1 General Mathematics	100	2.1 Materials	150	3.1 Mercantile Laws	50	4.1 Financial Mathematics	50	5.1 General Accounting	50	6.1 Financial Management	75
	1.2 Physics and Chemistry	50	2.2 Working Methods	75	3.2 Labour Laws	50	4.2 Statistics and Operative Research	100	5.2 Societies Accounting	50	6.2 Commercial Management	75
	1.3 Mechanics	50	2.3 Construction Methods	250	3.3 Fiscal Laws	50	4.3 Computers	100	5.3 Accounting Costs	50	6.3 Cost Management	100
	1.4 Technical Drawing	100	2.4 Installations	150	3.4 Urbanistic Laws	50			5.4 Accounting Analysis	50	6.4 Surveying and Budgeting Techniques	100
	1.5 Topography	75	2.5 Organization and Planning	150	3.5 European Community Directives	100						
	1.6 Graphic Repres. Methods	50	2.6 Execution Control	100								
	1.7 Common Law	75										
	1.8 Economic Theory and Enterprise Economy	100										
	1.9 Communications	100										
Total Times		700 h		875 h		300 h		250 h		200 h		350 h
TOTAL												2675 h

Fig. 14.1

14.2 PROFESSIONAL PROFILE OF A CONSTRUCTION
ECONOMIST

14.2.1 Professional intervention areas

1   Building promotion

1.1   Commercial function
   1.1.1 Market research
   1.1.2 Research and valuation of urban real state (building and land)
   1.1.3 Research and selection of distribution channels
   1.1.4 Research on promotion and publicity

1.2   Financial functions
   1.2.1 Estimating investment volume
   1.2.2 Profitability of investment projects
   1.2.3 Quantification and selection of financial resources

1.3   Administrative functions
   1.3.1 Design of structures for management organization
   1.3.2 Design of channels for communication and information
   1.3.3 Studies on personnel
   1.3.4 Management
   1.3.5 Selection of staff
   1.3.6 Information and performance control of staff
   1.3.7 Study and selection of a construction enterprise

2   Intervention on design drafting

2.1   Optimization and planning
   2.1.1 Collaboration in analysing and looking for constructive eco-
          nomic solutions
   2.1.2 Studies in timescales, costs and resources
   2.1.3 Translation of the studies on timescales, costs and resources
          into general and derived plans
   2.1.4 Stating execution plans

2.2   Valuation
   2.2.1 Defining and measuring work units
   2.2.2 Stating of elemental and unitary prices

2.2.3 Budgeting
2.2.4 Designing, updating price models

3   Intervention in work management

3.1   Planning
3.1.1 Control of plans and programmes
3.2   Valuing
3.2.1 Measuring of performed work units
3.2.2 Contradictory prices
3.2.3 Certifications

4   Intervention in construction enterprises

4.1   Production
4.1.1 Determination of optimum resources for optimum performance
4.1.2 Analysis of the inclusion of new works in the backlog of orders of the enterprise considering resources available
4.1.3 Optimization of general and assigned resources
4.1.4 Elaboration of plans
4.1.5 Investigation of work units prices
4.1.6 Works budgeting
4.1.7 Performance programme
4.1.8 Definition of stocks
4.1.9 Analysis of personnel
4.1.10 Economic influence of quality control
4.1.11 Optimization of equipment usage
4.1.12 Programming and planning control
4.1.13 Certification of works

4.2   Financial function
4.2.1 Analysis and determination of optimum cash flows
4.2.2 Analysis and determination of accurate financial resources
4.2.3 Analysis and determination of profitability

4.3   Administrative function
4.3.1 Determination of and accurate size of an enterprise in each situation
4.3.2 Design of an organization structure

4.3.3 Elaboration of information and communication system

4.3.4 Studies on personnel

4.3.5 Development of management

4.3.6 Legal matters (fiscal, mercantile, labour, administrative)

4.3.7 Defining policy and management of an enterprise

4.3.8 Development and interpretation of control systems

4.4 Commercial function

4.4.1 Guidance for action in public relations and closing-up to clients

4.4.2 Define tendering policy

4.4.3 Determination of offering prices and discussion of contractual bases

4.4.4 Definition of buying policy

4.4.5 Definition of subcontract policy

4.4.6 Selection of suppliers and subcontractors and discussion of contractual bases

4.4.7 Verification of purchases and subcontracts to production planning and programming

4.4.8 Following and control of purchases and subcontracts

4.4.9 Management of income and expenses or funding

14.3 GENERAL TRAINING PROGRAMME

14.3.1 Theoretical training

A detailed analysis of the proposed professional profile advised a division of the subjects under study into the following blocks:

1. basic training;
2. technology;
3. law;
4. specific mathematics;
5. accountancy;
6. construction economics.

Each block is developed further below.

(a) Basic training

The objective of basic training is to give the construction economist general knowledge that will be useful during a period of specific training. It is related to general science; graphic representation methods; principles of law, economics and communications. The areas proposed are:

1.1   general mathematics;
1.2   physics and chemistry;
1.3   mechanics;
1.4   technical drawing;
1.5   topography;
1.6   graphic representation methods;
1.7   common law;
1.8   economic theory and enterprise economy;
1.9   communications.

(b) Technological training

For studying construction costs in any of the various processes in which the construction industry is involved, thorough in-depth knowledge of the technology used by this industry is required. In order to succeed, the construction economist should have a broad knowledge of construction solutions, personnel management factors, control and planning. In this important stage of training the areas are:

2.1   materials;
2.2   working methods;
2.3   construction methods;
2.4   installations;
2.5   organization and planning tasks;
2.6   execution control.

(c) Legal training

The activity of building and construction companies is developed in a legal environment that has to be very well known for those accepting the responsibilities of managing management. It is useful for the construction economist to have some knowledge of the legal framework of his or her work. Some legal areas could be:

3.1   mercantile law;
3.2   labour law;
3.3   fiscal law;
3.4   planning law;
3.5   European Community Directives.

(d) Specific mathematical training

This area tries to cover a very important sector in the training of the construction economist. Proposed sub-areas are:

4.1   financial mathematics;
4.2   statistics and operational research;
4.3   computers.

(e) Accounting training

While the construction economist need not be an accounting expert, he or she needs sufficient information on accounting to formulate decisions concerning construction economics. The main areas are:

5.1   general accounting;
5.2   societies accounting;
5.3   costs accounting;
5.4   accounting analysis.

(f) Training on construction economics

In this field we should analyse problems and specific solutions related to some of the main functions inside a building or construction company in the study of the problems derived from taking decisions on investments or financing; in the development of marketing and commercialization and in the development of models and techniques of surveying and budgeting in the early stages of design, during its elaboration and during the execution of works. This area could be divided into:

6.1   financial management;
6.2   commercial management;
6.3   cost management;
6.4   surveying and budgeting techniques.

14.3.2 Practical supervised training

The activities of construction economists in every country and all the experience accumulated show that, as well as theoretical training, the future professional needs practical knowledge that improves and elaborates theoretical concepts. Nevertheless, working in a field does not guarantee that the experience agrees with the proposed goals.

So, this practical training should be supervised by a construction economist professional with a well-known reputation who will ensure that the directives of the training programme are followed.

14.4 TIMESCALES

There are a great variety of temporal programmes, with several justifications for each model.

We present a consolidated model, which we think could satisfy the requirements of society, which demands young people's participation. This point of view includes the contributions of the French delegation (CEEC Meeting in Seville, April 1990).

14.5 TIME ASSIGNMENT

The model that we are proposing is to provide a common body of theoretical training for all member countries, which will include specific matters in each country. Teaching time for each subject is divided up as follows.

1.   Basic training

1.1	General mathematics	100 h
1.2	Physics and chemistry	50 h
1.3	Mechanics	50 h
1.4	Technical drawing	100 h
1.5	Topography	75 h
1.6	Graphic representation methods	50 h
1.7	Common law	75 h
1.8	Economic theory and enterprise economy	100 h
1.9	Communications	100 h

TOTAL	700 h

2.   Technological training

2.1	Materials	150 h
2.2	Working methods	75 h
2.3	Construction methods	250 h
2.4	Installations	150 h
2.5	Organization and control	150 h
2.6	Execution planning	100 h

TOTAL	875 h

3.   Legal training

3.1	Mercantile law	50 h
3.2	Labour law	50 h
3.3	Fiscal laws	50 h
3.4	Planning law	50 h
3.5	European Community Directives	100 h

TOTAL	300 h

4.   Specific mathematical training

4.1	Financial mathematics	50 h
4.2	Statistics and operational research	100 h
4.3	Computers	100 h

TOTAL	250 h

5.   Accounting training

5.1	General accounting	50 h
5.2	Social accounting	50 h
5.3	Costs accounting	50 h

5.4	Accounting analysis	50 h

TOTAL	200 h

6. Construction economics

6.1	Financial management	75 h
6.2	Commercial management	75 h
6.3	Cost management	100 h
6.4	Surveying and budgeting techniques	100 h

TOTAL	350 h

Resume

1.	Basic training	700 h
2.	Technology	875 h
3.	Law	300 h
4.	Specific mathematics	250 h
5.	Accountancy	200 h
6.	Construction economics	350 h

TOTAL	2,675 h

15

Total quality management in construction

Michael Hartmann

15.1 MARKET CONDITIONS

In general the capacity of the European construction industry is bigger than the demand for construction – and this will be the case for years to come.

15.1.1 Competition in construction

For this reason alone competition is fierce. Too small a demand for construction increases interprofessional competition – if allowed by domestic rules. Until now, competition within construction has been mainly a domestic affair, but any imperfect markets are likely to be invaded. Interprofessional and transborder competition will accelerate when the EC Directives on Procurement of Construction, Procurement of Services and Construction Products have full impact on day-to-day business.

The very idea of the Single Market is to encourage free and fair competition – interprofessional and transborder. The only constraints are those established to protect consumers, workers on site and the environment. Under such market conditions satisfying the needs of the client – specified as well as implied – is imperative.

In the wake of weak demand for construction, the Single Market becoming a reality and clients articulating specifications of performance for use of a building for a period of 20–30 years, it is inevitable that the known services and patterns of cooperation will change considerably.

15.1.2 Needs of the client

The 'client' as one judicial person is becoming of less importance in the construction process. The client used to be the central *one* person, whose interests were in focus and who was capable of making all capital decisions. The traditional interests of the client have disintegrated into a multitude of interests – including those of the investor, the end user, and EC.

Until now the typical construction process has been initiated by a simple need for space for one purpose or another, combined with the financial means to start the process. Because of specialization within construction, overcapacity, and excess empty space in existing buildings, satisfying the investor has become the key to starting the process.

In a nutshell, any investor wants a sound prospect of getting a better payback from the investment in question than from any other investment. Until about five years ago investment in real estate was reasonably secure. The value of real estate was always going up – the only points of uncertainty were whether the situation was stable or whether a boom was either in progress or fading out. In future, money in the property business will in general not be generated from increases in value but from the margin between income and costs of running and maintenance. It is now possible for building owners actually to go bankrupt triggered by the general drop in the value of their property. Consequently the main issues for an investor considering whether the planned payback might materialize or not are the prospects of

* obtaining the income/rent needed;
* keeping the costs of running and maintenance at the planned level.

As property is seldom paid for with cash, the income to provide the payback depends on how attractive the property in question is in the eyes of those wanting to rent – over the entire period of investment.

The qualities in demand by end-users are:

* an attractive address;
* attractive architecture;
* the possibility of individual layout and decoration of space

To the end-user, an attractive address is usually in the centre of the city, or sometimes in nice, open country, but always well situated in relation to means of transportation. This is the reason why new buildings over railway junctions, in a harbour etc. or rehabilitation at the same places are so popular.

Second to the right location comes good architecture: nobody wants to live or work in a building of poor architectural quality. But the end-user's perception of architectural quality is in general rather different from that of the architects. Like the architects they want something 'special', but usability – functional and in terms of indoor climate – is never accepted as inferior to artistic virtues.

When the architecture is of satisfactory quality, the location is right and the price is reasonable, the possibilities of having the space custom-designed become important as a criterion for choosing between one offer and another.

15.1.3 Green issues

A subject of growing importance in construction is protection of the environment. At present, the legislation and rules concerning construction in most countries are very far from what is needed and what will be demanded in 5–10 years' time. In order to secure satisfactory income for the next 20–30 years, the design of today's buildings already has to be carried out with serious considerations on protection of the environment – dealing with choice of building materials, construction methods as well as running and maintenance.

15.1.4 Consumer protection

The most important issue regarding consumer protection is the possibility for the client to overview the consequences of signing contracts for construction. Because of national laws, the practice and procedure of contracts for design and construction are not at all clear to those unfamiliar with the subject. Some countries even have several kind of standard forms for agreement as well as general conditions.

In addition, many words acquire a different meaning if translated directly from one language to another. The differences in laws, rules, habits and terms represent the biggest obstruction to transborder trade, for the parties within the construction industry as well as the clients.

15.1.5 EC activities

The EC Commission first tried to pave the way for transborder trade in construction by way of harmonization. In 1987 the Commission engaged the French consultant C. Mathurin to map the national laws, rules and roles in the European construction industry. His findings were made available to the parties concerned in 1988, in what has become known as the Mathurin Report.

After mapping, Mathurin was engaged to compare the advantages and disadvantages of the various national systems, and to consider the possibilities of harmonization by way of EC Directives. The results of this second phase were published in 1989, and in 1990 the Commission finished its considerations on the possibilities of harmonization.

The lesson from the Mathurin study was that promotion of transborder trade in construction mainly by way of harmonization of laws and regulations was unlikely to be a success; it would need too many resources and be too slow to be of any interest.

15.1.6 Pan-European codes

The Commission decided that the best way to proceed was to encourage the pan European associations of the construction industry to develop codes of good practice, to replace the existing national systems. If such codes are developed to the satisfaction of the Commission as well as of the EC Parliament, the Commission will allow everyone acting according to such codes the right to use the CE mark for marketing their services.

15.1.7 Liability and cover

Only professional liability in construction and economical cover of liability will be regulated by an EC Directive. At first, construction was to be covered by the General Services Liability Directive, but after consultation with the pan European associations the Commission decided to commence preparatory work envisaging specific rules for liability in construction.

In a change from the traditional EC Commission working method, it was decided to involve the different construction-related professions and industry sectors right from the beginning, and so GAIPEC (Groupe des

Associations Interprofessionnelles Européennes de la Construction) came into existence.

The GAIPEC proposals for the EC Directive were delivered to the Commission in September 1992.

15.1.8 Demand for use of ISO 9001

All parties in the construction industry will in the very near future be forced to implement a quality management system based on ISO 9001: not because of EC directives or the like, but simply because their clients will insist. The demand for contracts to be based on ISO 9001 reflects primarily a determination to reduce the indirect costs for the client caused by defects – time spent on establishing proof of guilt, time wasted on misunderstandings, interprofessional friction, remedying defects and judicial muddle. When contracts are based on ISO 9001 it becomes simpler and quicker to establish proof of guilt than hitherto. A study by the Danish Ministry of Housing revealed that the costs of establishing proof of guilt in relation to construction defects exceeded the compensation obtained by way of lawsuits.

As the pattern in income from property is changing, the owners will stop accepting paying any direct costs for remedying defects. At the same time, taxpayers no longer want to pay for defects in public construction or subsidized housing. When the new EC Directive on liability in construction is enforced, the industry will have to pay for remedying defects. For this reason alone the parties in the industry need an instrument to minimize defects.

15.1.9 Information technology

At present the use of information technology (IT) in construction is in its infancy, but within five years or so all those who are still in business will make extensive use of IT – instead of the present fragmented use of isolated EDP systems and unintelligent exchange of data.

The use of IT is already or most certainly will be demanded by the property operators – those who on a business basis provide day-to-day management of property. But even before property operators demand the use of IT, competition will force everybody to use IT if they want to stay in business.

IT comes into its own when confronted simultaneously by price bidding, tight schedules and the demand for extensive documentation.

15.2 PRODUCTIVITY

Competition – further increased as the Single Market becomes a reality – will change the roles of the various professions as we have known them for some time.

15.2.1 Traditional roles

Traditionally, most construction start with a client consulting an architect. The architect then advises the client on whom to engage as surveyors, contractors etc. However, as the aim of construction is increasingly to make a profit mainly from the difference between income and costs, construction economics come into focus.

15.2.2 Construction economics

The construction economics in question are:

- feasibility studies on risk and the prospect of payback;
- cost information and timescales;
- project management, including negotiating contracts on behalf of the client with architects, engineers, contractors etc. plus planning and documentation of quality management;
- instructions on methods and information technology to be used for design and for capture of the data needed for documentation and planning of running and maintenance;
- briefing on design regarding usability, indoor climate, care for the environment, and security on site.

15.2.3 Client relations

The EC Directive on Public Procurement of Services will have a significant impact on relations between clients and consultants. There will be bidding for almost all kind of jobs. From now on, architects will not be able to act simultaneously as designer and as consultant to the client on which other consultants to engage.

15.2.4 Project management

The Services Directive institutes the role of the project manager as a function independent of and controlling design as well as construction and the collection of data needed for planning of running and maintenance – including documentation of quality management during the process from the very start.

15.2.5 Improvement of productivity

In order to attract investment in new construction projects, developers and project managers are forced to improve productivity substantially. The possibilities for increasing productivity are:

- improvement of planning – overall and for each activity;
- shopping for services;
- improving cooperation between the parties involved;
- minimizing defects in all activities.

Planning of design and construction is in general not carried out efficiently. As an illustration of the potential, it is worth mentioning a Danish study of logistics. On site, workers spend about 30% of their time in looking for materials, internal transportation on site and waiting for the components that for one or another reason have not been delivered on time. And this is in the context of the Danish construction industry, whose subcontractors are some of the most efficient in Europe.

Traditionally, shopping for services is only done by main contractors, who shop among subcontractors; this is not to the benefit of the client, or the subcontractors for that matter. Shopping directly among subcontractors can provide a cut in costs of 9–15% – minus the extra costs of planning and coordinating the subcontractors.

In general, calculated shopping among consultants is not common and certainly not welcome. All the same, clients as well as consultants might benefit *provided* the shopping is executed in a fair and rational way. To most consultants, well-organized shopping represents a possibility of levelling the workload and avoiding taking on the performance of services that the company or the employee in question is not fit to handle – professionally or with a satisfactory earning potential.

The Danish study on logistics in construction mentioned above revealed the possibility of a 10% cut in total costs if the entire process is planned to optimize interprofessional cooperation and logistics – including internal handling at the building merchants and direct delivery from the factory.

It is difficult to estimate the total costs of defects in the construction industry. However, some indication is given by the typical internal cost of subcontractors of about 10% of the contract sum – just to remedy shortcomings *before* handing over.

15.2.6 Project organization

To produce substantial improvement in productivity, construction projects have to be organized in a new way, allowing the project manager to plan and control the entire process as well as shopping for consultants and subcontractors.

Figures 15.1–15.3 set out a proposal for a new model for organization of project management.

15.3 QUALITY MANAGEMENT

The requirements specified in ISO 9001 are aiming primarily at preventing non-conformity at all stages, from design through to servicing. Dealing with quality management on the basis of ISO 9001 in construction presents two major dilemmas for the client, in domestic as well as in transborder operations:

- the risk that the project (the finished construction job) will not conform to either the stated or the implied needs of the client, even though each contract is fulfilled according to the contractual requirements (stated as well as implied);
- the lack of generally accepted procedures suitable as 'yardsticks' for planning, design and construction

Project development
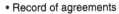

- Concept
- Letters of intent on sale or lease
- Agreement on investments

Preliminaries

- Record of agreements
- Specification of project
- Agreements on fact finding

- Agreement on feasibility studies

Preliminary planning

 • Collecting, checking and recording data

Feasibility studies

- Risk concerning time scales
- Risk concerning economics
- Meeting demand investment rate
- Functions
- Energy consumption
- Environmental issues
- Other studies

Production of brief

- Conclusion on feasibility studies
- Production of proposal for brief
- Checking review and completion of brief

Sales promotion

- Agreement on visualization
- Contracts on sale and lease
- Client's approval of further action

 • Visualization of project

Management of outline design

- Agreement on outline design including
  instruction on methods, H/W and S/W
  to be used

Outline design

- Proposals
- Production of drawings
- Production of specifications
- Production of timescales
- Budget
- Applications
- Checking data produced
- Handing over outline design

- Approval of outline design
- Client's approval of costs/time scales
- Client's approval of further action

Fig. 15.1

Brief on scheme and detail design

- Brief on performance requirements in terms of global costs, cleaning, energy consumption and saving, supply, recycling and disposal of water, interior climate, use of construction principles and products, considerations on the comfort of elderly and disabled etc

- Instruction on the use of methods, H/W and S/W

Management of scheme design

- Agreement on scheme design
- Design review

- Approval of scheme design
- Client's approval of costs/time scales
- Client's approval of further action

Scheme design

- Proposals
- Production of drawings
- Production of specifications
- Production of timescales
- Budget
- Applications
- Checking data produced
- Handing over scheme design

Management of detail design

- Agreement on detail design
- Design review

- Approval of detail design
- Client's approval of costs/time scales
- Client's approval of further action

Detail design

- Proposals
- Production of drawings
- Production of specifications
- Production of timescales
- Budget
- Applications
- Checking data produced
- Handing over scheme design

Fig. 15.2

Preliminaries to tender actions

- Proposal to client on site management
- Approval of concept of site management
- Agreement on planning of site

- Approval of planning of site

Planning of site

- Planning layout of site
- Specification of weather precautions
- Applications dealing with site

Tender actions

- Preparing tender actions
- Invitation for tender
- Answers to inquiries on tender
- Reception of and evaluation of tender
- Negotiations and acceptance of tender

Management of construction

- Agreement on site management
- Approval of contractor's design

Preparation of construction

- Design of workshop drawings
- Timescales for activities

Construction

- Approval of procedures for activities
- Management of construction and safety

- Authorization of an account payment
- Negotiation on costs of variation

- Supervision of reference activities
- Running inspection on site
- Checking documents of QM
- Monitoring of progress
- Acceptance of variations

Handing over

- Information to supervision on jobs finished
- Checking of jobs finished
- Collection of data needed for maintenance
- Checking of data for maintenance
- Acceptance of jobs

- Reception of data for maintenance
- Recording of acceptance
- Final payment to consultants and contractors and release of warranties

- One-year inspection

Running and maintenance

- Planning of day-to-day running
- Planning of preventive maintenance

- Five-year inspection

Fig. 15.3

15.3.1 Fulfilment of needs

These risks can be eliminated by organizing the project on the basis of the proposed new model – if the project manager is capable of the job. See Figs 15.1–15.3. In principle it does not matter whether the project manager is operating as a consultant or is employed by a design and build contractor. The crucial issue is whether or not he has the authority needed to execute the job properly. Contracts based on ISO 9001 provide the client ample protection only if the procedures in the quality management system correspond to the needs of the client.

15.3.2 Lack of procedures

From the client's point of view the most important issue in quality management is the way that decisions are made and the range of commitments consequent upon each decision. At present there is no pan-European code decision making and commitments – only various general conditions issued by the national associations of architects, surveyors, contractors etc. The proposal in Figs 15.1–15.3 represents the outline for a comprehensive decision model.

The other quality management procedures of importance in securing conformity at all stages, from the first idea to servicing after handing over, are:

- feasibility studies;
- the client's brief, including the identification and specification of implied needs;
- instruction on 'green' issues related to design;
- instruction on methods for design and use of software
- outline design;
- scheme design;
- detail design;
- planning of the site;
- instruction on design done by contractors;
- instruction on quality management for construction;
- instruction on weather precautions and safety on site;
- instruction on collection of data for planning of running and maintenance;
- tender actions;
- management of construction;

- handing over.

This chapter cannot describe in detail the content of all the procedures mentioned above. However, the following elements of project management and the production of instruction for methods of design are worth mentioning because they are essential to ensure conformity with the needs of the client as set out above in section 15.1.2:

- record of decisions and related documentation;
- design review and control of the validity of data;
- design aiming at minimizing life-cycle costs;
- identification and calculation of risk related to use of new methods of construction and new building materials.

15.3.3 Project record

Records of decisions made during the whole process, and of the material forming the basis of the various decisions are essential to document the liability of the parties involved and thus avoid waste of time on judicial muddle; see Figs 15.4–15.8.

15.3.4 Design review

Design review is a method for evaluating all the major choices made by the designers at the various stages of the design process – as soon as those choices can be illustrated and *before* further work is based on those choices.

Design review covers the evaluation of:

- architectural quality;
- location of functions;
- the individual functions;
- flexibility concerning layout of space;
- indoor climate;
- accessibility for elderly and disabled people;
- acoustics;
- daylight and lighting;
- fire precautions;

- security;
- cleaning;
- preventive maintenance;
- life-cycle costs;
- green issues.

At each stage – outline design, scheme design etc. – the design review deals only with subjects relevant to the stage in question. For that reason the agenda for design review is different from stage to stage. Figures 15.4–15.8 illustrate a selection of standard forms that I have developed for the documentation of design review.

15.3.5 Control of data validity

According to the 'right first time' concept it is feasible to implement working procedures that ensure the validity of all data when it is either imported or created in the design process. When procedures for control of data validity are developed and implemented, the individual has full responsibility – apart from auditing.

At the end of each stage – before activities related to the next stage are started – a global review of the data produced so far takes place. The purpose is to ensure that drawings, specifications, timescales etc. are sufficient and comprehensive.

15.3.6 Minimizing life-cycle costs

In order to keep within budget it is not uncommon in construction to accept solutions that have costly implications for running and maintenance. Often the client or subsequent owner does not realize the consequences until it is too late. As income in the future is likely to be generated mainly from the difference between income and the costs of running and preventive maintenance, this situation has to stop. Design review is a suitable tool for changing the situation.

The compulsory Danish Regulations on Quality Assurance (contractual demands for quality management) in construction now specify an obligation to design in a way that is economically favourable for running and preventive maintenance, based on an evaluation of life-cycle costs.

PROJECT RECORD

Project		Case no.	Document no.	Page

Item	Subject		Date	Appendix	Signature

Fig. 15.4

INCEPTION		IDENTIFICATION OF IMPLIED NEEDS		
Project		Case no.	Document no.	Page

The purpose of this action is to identify and specify implied needs and expectations as far as possible		Date
		Signature
Subject	Comments and specifications	Appendix

Fig. 15.5

OUTLINE DESIGN			RECORD OF REVIEW	
Project		Case no.	Document no.	Page

Participants in Review	Date
	Signature

Item	Subject	Results of Review	Appendix

Fig. 15.6

SPECIFICATION AND CALCULATION OF EXPERIMENTS

Project	Case no.	Document no.	Page

Description of the Experiment	Classification	Append.

Extent of the Experiment	Append.

Nature of the Experiment	Append.

Measures to minimize the Risk	Append.

Calculation of Risk in economical Terms	Append.

		Append.	Sign	Date
Yes/No	The experiment is approved by the Client			

Fig. 15.7

SPECIFICATION OF ADDITIONAL SERVICES

Name of project	Case no.	Document no.	Page

✔	Performance	Comments	Budget	Agreed
	Inception			
	Feasibility studies			
	Collection of basic data			
	Client's brief			
	Project manager			
	Quality management according to client's instructions			
	Specification of QM-instructions to contractors			
	Inquiry session			
	On-site management			
	Supervision			
	Inspection of starting procedures			
	Inspection of QM performed by contractors			
	Inspection of work			
	Inspection of data delivered by contractors			
	Revision of drawings etc. as built			
	Sorting of data for planning of running and maintenance			
	Services for planning of maintenance			
	Instructions for running and maintenance			
	Instructions to end users			
	Service after handing over			
		Total		

Agreement on fees

	Fees based on hours used		
	Fixed fees	Fees per hour	Total hours

		Reference	Initials	Date
	Budget for necessary hours, presented to client			
	Allocation of hours according to agreement			

Fig. 15.8

15.3.7 The use of new materials

Quite a number of defects are caused by solutions and building materials that have not been used before, or are not familiar to the contractors in question. The Danish QA Regulations also make it compulsory to obtain the client's acceptance of the use of new principles for construction or of new building materials.

As part of the procedure for acceptance of 'experiments' the designer has to describe the nature of the experiment, the risk, measures to prevent defects as a result of the experiment etc. Figure 15.7 shows a standard form to be used to document the specification of an 'experiment', measures to minimize risk, calculation of economical risk and acceptance of the client.

15.4 TOTAL QUALITY MANAGEMENT

The purpose of total quality management (TQM) is, in my view, to enable the consultants dealing with construction to face the challenges mentioned – and at the same time make a decent living.

15.4.1 The basic philosophy in TQM

It is the task of management to formulate and implement policies and procedures that enable all owners and employees to perform any activity correctly and rationally the first time – and, as an integrated part of activities, to store necessary documentation in the IT system.

15.4.2 Content of TQM

The elements of TQM are procedures for:

- rational production of services;
- internal planning of production;
- contact with clients;
- systematic collection and internal distribution of relevant professional knowledge;
- systematic collection and internal distribution of relevant experience.

At present, almost no services are produced on the basis of an intimate knowledge of what is needed and sufficient for the receiving part – including how the service must be structured and delivered in order to enable the particular recipient to consume the service in question with a minimum of effort. As an example can be mentioned the tender documentation, which often contains a lot of text not needed by the contractors and yet lacks important information, that is subsequently provided on request as the construction process progresses.

All jobs and contact with clients provide much valuable experience, but this information is rarely collected, sorted and distributed systematically for the benefit of new jobs, marketing, troubleshooting etc. – an enormous waste.

15.4.3 Use of information technology

Parallel implementation of TQM and IT is a huge challenge and resource consuming but at the same time presents a unique opportunity to review all procedures – not only those for production but also the interface with clients and the internal service backup system.

TQM may pay for quality management in production – and more besides.

16

Impact of information technology on the construction industry

Brian Atkin

16.1 INTRODUCTION

A single statement on the impact of information technology (IT) on the construction industry, within the limits of this chapter, would not only be difficult, it would also be bound to be wrong. Generalizations on the use of information technology are commonplace. Exceptions to the rule are, however, more likely to prove worthy of consideration, but may not always be given prominence. Too many writers and commentators give weight to what they see as the more important technological developments while ignoring the underlying issues. The result is that both the relative newcomer and the more experienced user of information technology may be none the wiser. Achieving the right balance may not be easy, but it is necessary. Attempting some kind of technological fix, through providing a so-called informed view of the latest solutions, would do nothing to advance our collective understanding of the issues involved.

This chapter, therefore, avoids promoting specific solutions or developments except where they are absolutely necessary to illustrate or help present a particular issue or line of argument. It would be unfortunate if they were seen as adopting the easy option. The alternative might have been to produce, say, a catalogue of computer usage. While this might make some people feel pleased with themselves if they were shown in a good light, it would hardly advance our knowledge nor would it instil confidence in our clients.

It is also probably fair to say that a conference that promotes construction economics, or rather the work of the construction economist, would want to hear that the profession's use of information technology is well advanced and helping to provide clients with the answers they are seeking. So it is; but this would be too glib a statement to make and one that surely could not stand up to close scrutiny. Criticizing the present use of information technology may be fashionable, but is also predictable in what it will say. The issue then is one of how to inform a readership that has mixed experiences and expectations? An equally difficult task is defining information technology to everyone's reasonable satisfaction; after all, the construction industry is not a homogeneous group.

For this author, the primary task is to communicate to a readership that, at all levels of experience, might benefit in some way from what he has to say. In the first instance, however, information technology must be defined. My view is that we should not be dogmatic about its exact extent and meaning. 'Computer-based technology that assists in transmitting, manipulating, storing and presenting information' will surely suffice.

16.2 WHAT HAS INFORMATION TECHNOLOGY ACHIEVED?

The 'impact of information technology' can be interpreted in more than one way. It was once seen as a threat to the preservation of a way of life (or work) that had for so long seemed immune from major change. Very little evidence can be produced to substantiate claims that information technology – computers in this sense – has put professional people out of work, at least not yet. True enough, the careful use of technology can now increase an organization's capacity for work without a commensurate increase in personnel.

Technology promises many things, yet progress in absorbing information technology in construction has, in comparison with other industries, lagged behind. This may be about to change, if evidence coming to the attention of the Commission of the EC is to be believed. Details of a major study have begun to emerge through reports in the technical press. According to those reports, the role of the designer *will* change in a number of ways. For instance, designers will leave much of the detail design for a building to specialist trades' contractors and manufacturers. Additionally, the person at the centre of the design and management of a building project may no longer be an architect or engineer, but an information manager (see later section for a discussion on this topic). This development will have implications for the future role of architects, especially as information technology is likely to hasten the erosion of their traditional role. CAD systems and other systems that can interact with them are not exclusive to architects or engineers. Other construction professionals are likely to see them as a means by which they may extend their own influence. Whether or not this will lead to direct improvements in internal efficiency within the industry remains to be seen. What is clear is that new, competing pressures will come to bear on the design and construction team, with the effect of rede-fining roles and responsibilities.

A simple view is that technology provides us with the solutions, but first we must know what the problems are and be capable of explaining them to others. For too long, many of the problems that are solved in the course of a construction project have been regarded as unique, or very nearly unique, to construction. A significant problem has been in perpetuating the view that each building project is different from the next. This may well be so, but there are clearly similarities in design features and in the management skills required. So why then do we waste resources by reinventing prob-lems time and again? Having an industry that is flexible means that a wide range of building projects can be mounted, with demand for basic resources fluctuating according to the requirements of the marketplace. Information technology could, for instance, be used to greater effect in storing our knowledge of successful problem-solving and in using that knowledge to help reduce the risk for clients on future projects.

16.3 PROBLEMS AND ISSUES

Construction economists do not work in isolation; they are part of a team effort that can and does deliver outstanding buildings. Sometimes the results exceed expectations. Designing and constructing a great building like La Grande Arche or Canary Wharf demands the highest level of pro-fessionalism. But these projects represent the leading edge and are not typi-cal of the vast number of buildings that are designed and constructed for more routine or ordinary uses. Greater success in exploiting information technology, for the benefit of serving owners of the majority of buildings, will stem from better understanding of the information and communication systems that both support and nurture the design and construction process.

This need is readily apparent if we look at the extent to which informa-tion technology, or more specifically, computers are used by construction economists and indeed other construction professionals. Few if any organizations could boast of having a total system of data gathering, manipulating, storing and reporting that provided a seamless interface between all the stages or phases in a building project. While there may well be excellent examples of computer-based systems, which can show demonstrable improvements in working over manual systems, they are more likely to be operating in isolation and not as an integral part of a total system.

So why has information technology lagged in construction? Finding the real answer to this question might help us to form a better view of the most promising development paths. By understanding the real barriers to communication we might reach some better answers.

Let us start by looking at the characteristic nature of the construction industry. In most countries it is an unregulated, highly fragmented industry where market forces dictate who gets the work. The use of information technology has been seen by some organizations in the marketplace as the key to competitive advantage. Even so, these organizations might argue that they are prevented from maximizing their advantage because of inherent defects and inefficiencies in the way the industry operates. A simple example would be in designers' use of computer-aided design (CAD) systems. Here, there is an opportunity to provide the construction economist with improved and more rapid information on a design. Failure to do so can be blamed on a rewards system that reinforces traditional practices and values.

In other words, designers will not and cannot be expected to do more work for which they see no direct benefit to themselves. This is called commercial realism. Ponder then on an information and communication system that enabled more complete and meaningful data to pass from designers' CAD systems to the construction economist and others. Not only would the overall time of a project be reduced, it might also lead to an improvement in quality, as a situation where communication is rapid leads to an increase in iterations between design generation and analysis. Put simply, there is more scope for testing design options which, in turn, might lead to the establishment of a better overall solution.

16.4 INFORMATION MANAGEMENT

Flows of information are crucial to the success of a building project. Too much information and the project becomes swamped; too little and it becomes starved. Information is the life-blood of any project, for which accuracy, timeliness and completeness are important factors in determining

its health. In other industries where major capital projects are the norm –
shipbuilding, aerospace, oil etc. – information management is regarded as
an essential function. Unlike project management, its role is not strategic or
tactical, but plainly operational. Ensuring that the right information gets to
the right person at the right time in the right format are fundamental objectives of any information management system. It is part of the supporting
system or infrastructure of a project. Architects, engineers, construction
economists, construction managers and so on constitute the professional
team that conceives, details and manages a project. The fruits of their
labours are not so tangible as the bricks and mortar of the building, though
they are no less important. Information, be it paper-based or computer-
based, is the measurable output, its quality having some bearing on the
eventual quality of the building (Table 16.1).

Table 16.1 Information volume by contract value – an indication of the size of the
problem

	Contract value				
	£350M	*£40M*	*£25M*	*£7M*	*£5M*
Contracts	300	80	170	100	50
Tenders	1,200	300	700	400	200
Drawings	5,000	1,000	30,000	4,000	200
Drawings issued	400,000	100,000	150,000	48,000	30,000
Variations	10,000	1,000	5,000	150	300
Site instructions	30,000	3,000	6,000	600	1,000
Rooms/areas	3,000	100	11	7	50
Consultants	15	8	11	12	5
Cashflow/month	£7M	£1M	£0.75M	£0.3M	£0.25M
Approvals/week	70	26	n/a	n/a	13
Meetings/week	20	10	30	11	6

As outlined earlier in this chapter, reports are emerging of a major study
that brings into focus the role of the information manager. Let us assume,
therefore, that the case can be proven for a distinct information management system on each project. Who then will drive it? Look no further than
the construction economist. Sitting at the centre of the project, the con-

struction economist commands the position that is best suited to organizing and managing the flows of information. Further proof of this assertion can be seen in the nature of the construction economist's work, where value is added to the information that flows into and then out of the organization.

A number of developments, one commercial (Crow–Maunsell) and several created within and for the organizations concerned, show what can be done. The common feature of these systems is that they are centred on a relational database structure. This holds interesting possibilities, as a progressive move towards object-oriented designs – as supported by today's CAD systems – would certainly be compatible with this approach. In any event, however, we may not be thinking of a single, physical database, but several, linked by a common product data classification system or PDI (product data interchange). Work has, of course, been done to evolve the international standard, STEP, though far more needs to be done to bring fully operational systems, supporting STEP, into routine use. It is unlikely that there will be serious challengers to STEP and so we shall eagerly await its arrival.

16.5 ELECTRONIC DATA TRANSFER

There has been firm interest in electronic data transfer or interchange (EDI) for many years, though few tangible successes have occurred. The UK's EDICON group was formed in 1987 and its role model has spread to nine other European countries. EDIBUILD has now been formed as a pan-European group to foster a common approach to the development and implementation of EDI through the construction and related industries.

The electronic exchange of bills of quantities between the building owner's construction economist and a construction company has recently taken place successfully. Monk Dunstone Associates have exchanged sections of a bill, using the EDIFACT BQ message structure developed by EDICON, with one of the UK's leading construction companies, John Laing. Having received the bill, Laing were able to download it to their own system to create an estimate that was then returned, in the form of a priced bill, to Monk Dunstone using the same EDI message format.

Other message formats that have been established to date include invitations to tender, tendering, establishing the contract, quantity valuation, payment valuation and direct payment valuation. Perhaps surprisingly, message formats for graphical data exchanges (i.e. through CAD) are also being developed.

A look at developments worldwide would suggest that Western Europe is ahead of the field. The Message Development Group (5) or MD5 for short is well established, drawing representation in from the UK, France, Germany, Netherlands, Denmark, Switzerland and Finland: all these countries are, by happy coincidence, active members of CEEC. EDIBAU of Germany is involved in developing the CONWQD (Construction Work Item Determination) message format. The aim is to detail the individual components of work that may be incorporated in composite form within the bills of quantities. COMBINE, a French system developed by CSTB, is focused on information exchange between tools for evaluating performance. And there are other initiatives like DECIDE-S, which seems to be gathering pace in Sweden and Denmark.

16.6 SUMMARY OF A STUDY

This section summarizes the results of a study (Atkin and Miller, 1993) aimed at establishing both the current and the likely future use of information technology by construction professionals in the UK. The study is especially timely as it provides some insight into the way in which the severe recession in the UK has affected information technology implementation strategies. The results have also proved illuminating in that they have confirmed some of the trends discussed earlier in this chapter.

The study was carried out in two parts: first, a questionnaire was designed and tested on a pilot group, then released to a sample of organizations, large and small, that were known or believed to be active users of information technology. A total of 128 usable responses were received. Subsequent interviews with a select number of respondents were used to flesh out further information on, for example, attitudes, strategies and the potential for applying information technology.

Respondents were asked to consider several reasons for their use of information technology and to rank them accordingly. Most respondents saw information technology as a natural part of the equipment provision in their offices, ranking this the highest. Of the other reasons considered, the use of the technology to add value to service was ranked lowest by most respondents. These, and the remaining rankings, would appear to suggest that information technology is seen as tool for cutting costs rather than as a means for adding value to service. The notable exceptions were several large construction economist (quantity surveying) organizations, where their emerging role as information managers was seen as helping to add value to the information and services they provided. Progress is, however, impaired: not simply because of recession, but arising from the difficulties of electronic data transfer. Until workable standards emerge and are embodied in robust systems, there is unlikely to be any major improvement in data transfer. Respondents and interviewees recognized the strategic importance of EDI and STEP for the industry, but felt too far removed from what was going on – this was especially so in the case of STEP.

16.7 CONCLUSIONS

This chapter was not conceived with the aim of supplying statistical evidence of the use (or non-use) of information technology in the construction industry and, in particular, in the work of the construction economist. The author might be criticized for not providing comparative data on the use of the technology and the types of computer-based systems involved. Such data are available, but devoting this chapter to describing them would not help to move things forward. If data alone has the potential to move practice on, why then has this not occurred in the past? Furthermore, there was, in the view of the author, little point in citing CAD as the technology of the future: it is a technology of the present and provides increasingly powerful solutions to designers' needs. Expert systems have been developed and shown their potential and, in one or two cases, work well in narrow domains within current practice.

The choice of topic for discussion in this chapter was centred on two key developments. One was driven by commercial expediency: that is, how can transactions between multiple parties to a project be speeded up? The other arose for mainly pragmatic reasons: that is, the need for the efficient and effective management of that information. In both of these areas, the construction economist has a vital, if not central, role to play. To be successful at both, the construction economist must develop an even broader view of the construction industry and the ways in which it serves its clients. Through this can come real strength and influence, as well as a more certain future.

REFERENCE

Atkin, B.L. and Miller, S.A. (1993) *A Study of Information Technology in the UK Construction Industry*, Summary Report, University of Reading, UK.

17

Risk management and insurance

Bjørn Roepstorff

17.1 INTRODUCTION

The purpose of this chapter is to highlight the importance of risk management in general terms and the use of this concept within construction in
particular. Further, some aspects of insurance in this context will be dealt
with, seen partly from a professional indemnity point of view and partly as
an important tool to finance losses encountered by all parties in the construction of a given project.

Risk management may be defined as a 'systematic process of identifying,
measuring, directing and financially controlling risks threatening a client's
assets, liabilities and future earning capabilities'.

One could say that the real objective of risk management is to reduce fear
of the unknown and the unexpected, and to create confidence in the future.
As such we are dealing with a very old discipline, which has been developed into modern management theory in the USA in the 1950s and 1960s
and still is under development in the 1990s.

17.2 FUNDAMENTALS OF RISK MANAGEMENT

The fundamentals of risk management may briefly be divided into the following categories.

- Risk analysis:
  - identification (threats, key areas);

- – measurement (quantification, qualification);
  - – risk profile (priorities).
- Risk evaluation (risk trade-offs based on cost/benefit, acceptability, social/political considerations):
  - – avoid;
  - – eliminate;
  - – reduce;
  - – transfer;
  - – retain.
- Risk handling:
  - – change of risk (physical, immaterial);
  - – risk financing (retention, insurance, captive insurance)

The elements of risk management can be regarded as a paternoster: that is, as a continuous task throughout the project, which implies coordination, communication, consistency and lessons to learn (education).

A successful implementation of risk management depends on a written and clear **risk management policy**. This policy must clearly describe the main goal of the project, the strategic considerations, the financial strength and the willingness to assume which risks of the project.

17.3 RISK ANALYSIS

For a construction economist the aim of risk analysis generally could be described as being:

- to collect and handle information to determine a project's **risk profile** in order to reveal whether it lives a 'natural' life of risk;
- to give the owner/management a better foundation to evaluate proposals for changes of the risk profile, including proposals for elimination, reduction, transfer or retention of risks; and
- to secure continuous information about possible changes in the risk profile and the causes of these changes.

There is no standard formula for risk analysis; however, there are several recognized techniques and sub-techniques in various areas.

The **engineering** type of risk analysis is systematic, and implies a quantification of risk levels based primarily on technological considerations. These are quite well developed through the years. The EEC Risk Directive of 1982, which regulates the relations in connection with major accidents in industrial activities, clearly demands that the existing risks for major accidents are explained and evaluated through an engineering type of risk analysis.

Other types of risk analysis are the **decision–analytical** method (multidimensional optimization based on mathematical models); the **risk-perception** method (seek to understand why people react differently from what the theoretical analysis predicts); and the **political–analytical** method (seek to explain how social and 'political' attitudes influences the design and implementation of the risk policy).

Finally, the **contract–analytical** method should be mentioned. This method seeks to clarify the risk positions of the parties involved and to use this knowledge actively in order to improve the risk position of the client by recommended strategic measures.

No single method fulfils the previously stated aim of risk analysis; rather, a combination of two or more methods comes closer to the true picture.

The following **rules of thumb** can be given for working with risk analysis.

- **Define** in advance which results may be described as disadvantageous or advantageous (subjective).
- **Choose** which factors (health, ecological, economic, social, political) should have the highest and subsequent priorities in the analysis.
- **Evaluate** the size of the damage to which the environment and the project may be exposed.
- As far as possible **reckon** the likelihood of different outcomes.
- **Decide** who will be affected by which hazards/risks, including any variation in time.

17.4 COST OF RISK

In a project you will as a construction economist discover that a calculation of the cost of risk for the project is not only a valuable exercise but also an essential part of your consulting work to your client, especially when alternative solutions have to be considered.

The cost of risk that ultimately has to be borne by a project may consist of the following items:

1. **insurance and guarantee premiums**: e.g. fire, storm, liability, transport, workmen's compensation, automobiles, trucks, cranes, lifts etc., bid bonds and performance bonds;
2. **claims prevention costs**: e.g. fire prevention, security systems, protection of the environment, evacuation planning and quality assurance;
3. **risk allocation costs**: e.g. market research, risk analysis, legal and auditing fees;
4. **administration**: e.g. risk control and insurance;
5. **risk retention**: e.g. deductibles/excess, self-insurance, non-indemnified claims committed by contractual partners or third party.

'Cost of risk' calculated as a percentage of project value or revenues is used partly as an internal and partly as an external benchmark for performance measurement of the risk management function.

17.5 COMPARISON OF RISKS

As risks are interrelated, e.g. energy production facilities, environmental impacts etc., we constantly in one way or the other in the future have to make **risk trade-offs**. To illustrate what is meant by risk trade-off consider the following example.

A warehouse of 8,600 m² with high fire loading is planned to be built. The first step could be to determine how many fire sections we need, by looking at the matrix in Table 17.1, which shows what we should call **technical exchange of risk**, indicating the normal applicable maximum size of fire cell in m² for different fire loadings.

Table 17.1 Technical exchange of risk

	Fire loading		
Fire-reducing measures	*Low*	*Medium*	*High*
No measures	2,500 m²	1,200 m²	600 m²
Fire ventilation	5,000 m²	2,500 m²	1,200 m²
Fire ventilation and automatic fire alarm	10,000 m²	5,000 m²	2,500 m²
Fire ventilation and sprinkler	Unlimited	Unlimited	Unlimited

If only fire ventilation is installed, we need in our case to have eight fire sections or cells compared with only one unsectioned area of 8,600 m² if we choose fire ventilation plus sprinklers. Assuming we have chosen the latter option, we can start making our cost–benefit calculations by finding the net investment amount:

Investment calculation

		DKK 1,000	DKK 1,000
1.	Total costs of sprinklers (3,000)		2,000
2.	Savings		
	Fire ventilation 2 → 1%	250	
	Simplified building construction (light steel construction, light roof)	500	
	Fire cell 3 → 1	500	
	Smoke detectors	30	
	Less waste space (4%)	520	1,800
3.	Net investment for sprinklers		200

Further, we shall find the effects of this net investment on the annual costs of running the building:

Annual costs

		DKK 1,000	DKK 1,000
1.	Increase		
	Inspection, maintenance	30	
	Depreciation (20 years)	10	
	Interest	20	50
2.	Decrease		
	Insurance premium	100	
	Maintenance of fire gateways	20	
	Wages of truck driver	25	145
3.	Annual reduction in costs		95

The outcome of these calculations gives a payback time of 2.1 years (200:95), which is not too bad. The traditional way of calculating the benefits of introducing risk-reducing measures often produces poor results. Further, the savings 'hidden costs' in the case of damages are often forgotten.

We have constantly to confront and compare risks. We can only do this in a professional way if in our risk assessment we have interpreted the risks broadly enough, and do not merely displace but substitute one risk for another: this can be a long and arduous job if done properly.

Which risk is your client willing to run in order to achieve a given goal? This depends on:

- the magnitude of the risk in money terms;
- the likelihood of the occurrence of the risk combined with the amount of the risk;
- the decision-maker's personal attitude to risk;
- what is actually achieved by the given goal.

17.6 CODE OF PROFESSIONAL PRACTICE ON RISK ISSUES

Effective as from 1 March 1993, a Code of Professional Practice has been established under the provisions of the Engineering Council's Royal Charter for the benefit of 290,000 registered engineers and technicians in the UK. This code aims to encourage greater awareness, understanding and effective management of risk issues.

The ten-point code with explanatory notes is shown below. The advice given could apply to any construction economist or anyone involved in risk management.

1. **Exercise reasonable professional skill and care.**
   You have a responsibility to exercise reasonable professional skill and care in the performance of your work. You have a particular responsibility when forming a judgement about the tolerability of risk.

2. **Know about and comply with the law.**
   Keep yourself up to date with the substance and intent of the legal and regulatory framework that applies to your work. Act at all times in a manner that gives full effect to your obligations under the law and the regulatory framework. Seek professional advice at an early stage if you

have any doubts about the appropriate application of the law or regulations.

3. **Act in accordance with the codes of conduct.**
   Familiarize yourself with the Engineering Council's Code of Conduct, authorized under its bye-laws, and with any relevant codes provided by your own institution or professional association. Act at all times in accordance with the requirements of the appropriate codes of conduct, and recognize that your broader responsibility to society may have to prevail over your personal interests. Respect your duty of confidentiality to your employer or client, and follow the appropriate procedures within your organization for raising concerns about potential hazards or risks.

4. **Take a systematic approach to risk issues.**
   Risk management should be an integral part of all aspects of engineering activity. It should be conducted systematically and be auditable. Look for potential hazards, failures and risk associated with your field of work or workplace, and seek to ensure that they are appropriately addressed. Balance reliance on codes of practice with project-specific risk assessment; be open-minded and do not hide behind regulations. Do not exceed your level of competence on risk issues or ask others to do so; seek expert assistance where necessary.

5. **Use professional judgement and experience.**
   Judgement is required to match the approach to the nature of the hazard and the level of risk. This might vary from a simple assessment to a formal safety case. Uncertainty is a feature of many aspects of risk management. Be aware of this, and use risk assessment methods as an aid to judgement, not as a substitute for it.

6. **Communicate within your organization.**
   Communicate effectively with colleagues, both up and down the chain of responsibility, to help ensure that risk management activities are sufficiently comprehensive and understood. Endeavour to raise awareness of potential hazards and risk issues among your colleagues. – Seek to ensure that all those involved with a project are aware of any risks to which they may be exposed, of any relevant limitations inherent in the design or operating procedures, and of any implications for their con-

duct. Discuss the reasons for incidents and near-misses with your colleagues, so that the lessons can be learned.

7. **Contribute effectively to corporate risk management.**
   Help to promote a culture within your organization which strives for continuous improvement, securing involvement and participation in risk management at all levels. Give due attention to risk analysis, evaluation, decision making, implementation and monitoring during all phases of an engineering project to ensure effective management of risk. Seek to ensure that management systems do not allow risk issues to be ignored, subverted or delegated to levels which have no control. Consider the cost implications of all aspects of risk management.

8. **Assess the risk implications of alternatives.**
   Always consider the possibility of risk reduction or avoiding a source of risk completely. The tolerability of risk will vary with context, and the basis for establishing it needs to be understood. In determining the tolerability or otherwise of a given risk, promote effective consultation with those who may be exposed, where this is practical. Some risks are so great that they cannot be tolerated under any circumstances, while others are so low that they can be tolerated without further justification; between these extremes, assessment is needed.

9. **Keep up to date by seeking education and training.**
   Risk issues and approaches to risk management should be integrated into every engineer's initial education and training. As part of your continuing professional development, seek education and training in risk management techniques. Increase your awareness of the range of potential hazards and learn from past events.

10. **Encourage public understanding of risk issues.**
    Contribute to the education of the public where you have the opportunity, so that they can be aware of and form an objective and informed view on major risk issues. Seek to encourage a positive public perception of the engineer's role in the management of risk. Contribute to improved communication on risk issues between your organization and the community.' (Source: *Foresight*, December 1992.)

17.7 ASPECTS OF INSURANCE

One of the important methods within risk management for reducing fear of the unknown and the unexpected, and to be confident of the future, is **to transfer the risk** to somebody else by contract. I am here thinking of insurance.

Many of you would agree with the saying, 'Things do not cause losses, people do'. Therefore it is necessary for insurers to weigh a given risk not only against objective criteria, but also and equally against subjective criteria.

Insurers are suffering increasingly from declining profits and solvency margins and are making great efforts to remain competitive and improve their profits. In the long term the insurance business has to be profitable. In fact, it has to be a bit more than profitable because the insurers have to meet their obligations at any time.

The larger buyers of insurance have other needs of insurance, due to their financial strength alone, than smaller buyers. Further, larger buyers are not prepared to pay the same premiums as smaller buyers.

17.7.1 Transaction costs

On average, for every FF100 paid in premiums, only about FF70 is eventually returned to the customer as payments of claims. From some insurers or in some fields of insurance you will only get back 30–50%.

A recent survey in the USA found that the industry-average expense ratio for 1991 was 26.5%:

	%
Commissions/brokerage	11.1
Salaries	5.9
Advertising	0.2
Boards and surveys	0.9
Taxes, licences and fees	3.4
Insurance	0.1
Travel	0.4
Rent	0.8
Equipment	0.9
Print and stationery	0.3
Post and phone	0.5

Legal and auditing	0.3
Other	1.7
Total	26.5

17.7.2 Captive insurance

The size of these transaction costs is not acceptable to the larger buyers of insurance. Consequently, they either negotiate better terms or form their own insurance company: a so-called **captive** insurance company. The success of captive insurance companies in general is partly due to the fact that they are able to reduce these transaction costs drastically compared with traditional insurance market placements.

By using a captive solution, around 90–95% of the premium could be available to pay claims instead of 70–75%. Obviously, in the present very competitive insurance market the withdrawal of the relatively 'good risks' from the traditional market will leave the insurance industry with correspondingly higher transaction costs. Therefore we are now seeing not only dramatic cut-backs in personnel but also an increasing premium level where it does not hurt too much. The 'poor' or weaker buyers have to pay for this. From a marketing point of view the insurance industry generally could be described as an inefficient business, which has survived only due to the absence of qualified counterparts.

In my opinion the most advantageous feature of establishing a captive has to be seen from a risk management point of view. No doubt the company will be more risk-conscious, which should eventually lead overall to lower cost of risk as formerly described.

However, there will always be a market for insurance. The catastrophic risks are best covered by insurance, and as economic decision-makers are risk-averse they are prepared to accept steadily increasing transaction costs as described.

17.7.3 Professional negligence

Operating as a construction economist and a human being at the same time means that there is an inherent hazard of professional negligence. This term may be defined as 'such a neglect of professional duty of care as to render the professional person committing the act, error or omission of neglect liable in law to a client or some other third party who occasions loss by reason of that neglect'.

17.7.4 Professional indemnity insurance

In Denmark and other Nordic countries there is a long tradition for having a professional indemnity insurance, normally negotiated with the insurance market by the Professional Organization of Architects. If employed by a firm of architects the services rendered by a construction economist are covered by the firm's normal professional indemnity insurance. The premium, however, has been the normal one for architects, even though we are talking about a very different type of risk.

The needs of insurance cover are different from the traditional ones. It is my personal opinion that these needs should be scrutinized and quantified/qualified so that the final risk description will attract the insurance market in such a way that the best possible conditions and terms can be obtained.

17.7.5 Construction insurance

At the planning stage of a building or project the building owner has so many important tasks to do that some of those deliberately or unconsciously get downgraded. A dangerous area to downgrade is insurance, but this is what often happens.

In most countries by law or by contract the owner of building has the duty to take out certain classes of insurance for the whole project. But fulfilling these legal obligations barely covers the need for insurance.

The great gaps in cover between the parties involved – building owner, architect, engineer, contractor, subcontractors and suppliers – can often create disputes, which produce delays, bad workmanship and losses due to uncovered claims.

A professional way to do it is to take out a **contractor's all risks (CAR)** policy, which as an umbrella covers the building project and the gaps and overlaps between the insurance cover taken out by the various parties involved in the building project, but does not replace them (e.g. professional indemnity, theft, transportation, employers' liability, storm and water damage). If the construction work includes erection of machinery the insurance has to be extended or a separate erection all risks (EAR) has to be issued.

The CAR insurance consists of two main sections: all risks cover and liability cover. The following sections provide brief notes on these forms of cover, as normally available on the market.

17.7.6 All risks cover

This section may cover a long list of risks either as standard or by extension (ext.) in connection with:

- the contract work itself;
- the building owner's existing buildings (ext.);
- machinery, material, tools etc. (ext.);
- clearing costs (ext.).

Who?
Building owner, contractor and their subcontractors.

Where?
On site, but may be extended to cover building material during domestic transport to site.

What?
Typical losses of or damage to the insured objects:

- storm;
- waterspout;
- collapse;
- theft on site;
- malicious damage;
- floor and other water damage;
- vehicle impact;
- subsidence;
- damages caused by work tools.

If extended, also covered:

- clearing costs;

- additional costs, e.g. overtime;

and if erection of machinery is included also:

- engine breakdown;
- test load;
- test run;
- commissioning.

When?

From the start of the contract until the project is delivered or starts to be used by the building owner and – if agreed – for the maintenance and guarantee period, usually 12 months up to 5 years (10 years).

Which?

Apart from the actual building and works, and possibly machinery, the insurance may (if agreed) comprise:

- existing buildings and works belonging to the building owner;
- machinery, material, tools, sheds etc. belonging to the contractor;
- clearing costs.

Exceptions?

The most important exceptions are:

- fire damage (covered separately);
- normal relief for shortcomings and defects;
- penalty for delay, non-completion or other non-fulfilment of contractual obligations;
- loss of profits of any kind;
- loss or damage caused by:
  wear and tear, and gradual deterioration;
  mechanical and electrical interruption or disturbance;
  faulty workmanship;
  defective materials;
  defective projection, calculation and construction.

However, consequential damages derived from the latter are covered.

17.7.7 Liability cover

The liability cover may be written solely for the building owner, solely for the contractor, or for both.

The reason for this cover is that the contractor's employer liability insurance often contains various exclusions that the CAR does not have, e.g. digging up, demolition, blasting, piling, bunging, lowering of groundwater and underpinning.

What?

During the period of contract work the liability for property damage incurred by the insured: i.e. building owner, contractor or subcontractors. Liability for bodily injury incurred by the building owner is covered as well.

Further, the liability for tortious consequences that arise after the completion and delivery of the work, provided they are due to errors or omissions committed during the period of construction. Such claims to be reported by the end of the first year after completion of the period of construction at the latest.

Exceptions?

The most important exceptions are liability:

- as a consequence of agreed or incurred liability that is more extensive than the general legal rules dictate;
- for damage caused by dogs, aircraft, ships or motor vehicles unless the latter is used as a work tool on site;
- for damage to property that belongs to, or is borrowed or rented by the insured or in another way kept in his custody;
- for damages to work object or part thereof;
- for losses due to non-compliance or in any other way non-contractual.
- for damage that is or would be covered under the employers' liability insurance or any other insurance.

17.8 CONCLUDING REMARKS

I hope that this brief survey of the general concept of risk management leaves a feeling that this can be much more integrated in the services rendered by the construction economist, not only for the benefit of your client but for yourself.

Some readers will be familiar with the insurance aspect of risk management, but I am convinced that there is a general need for knowledge of insurance so that this area can be more operational as a natural part of the construction economist's services to the client as well.

Of all catastrophes it is only the natural ones (elemental perils) for which you cannot blame the human being. In up to 90% of cases the cause has been human behaviour: e.g. lack of prediction, misunderstanding, lack of caution or ignorance on the part of management or employee. The number of mistakes that are made in a project or company is a function of the culture of the project or company.

Consequently, risk management should in future deal more with people than with property and liability, because it is people that constitute the biggest risk factor of business life.

I should like to finish by summarizing my experience in risk management in the following rules of thumb:

- Evaluate the risk.
- Fix the maximum possible loss.
- Do not risk more than you can afford to lose.
- Do not risk 'too much' in exchange for a chance for 'too little' gain.
- Remember to improve the risk into your favour.
- All the time be prepared to identify and exploit genuine chances (risk possibilities).
- Keep the broad view and keep a buffer in all essential matters.

18

The environmental equation: the impact of green issues on construction

T.D. Burton and C.F. Stoker

18.1 INTRODUCTION

It is now around 20 years since the first major international energy crisis took place. Although this did not last long, the crisis led many countries to develop national energy plans. At the end of the 1980s the extent of the environmental problems that we face came ever more sharply into focus. The problems were – and are – more serious than expected.

As an introduction to their National Environmental Policy Plan in 1989 the Dutch Minister of the Environment wrote:

> The environment is in a very critical condition. In spite of improvements in some areas, the situation is continuing to deteriorate. Further postponement of drastic measures is unjustified. Radical decisions will affect everyone and are unavoidable. The coming years will be characterized by a hard struggle; and it is not only the improvement of the quality of the environment which will be at stake, but ultimately the continued existence of mankind!

Dramatic words, painting a dramatic picture. The message is clear. It is going to affect all of us. We are beginning a new era where energy-conscious and environmentally aware building and living will become more than a hobby, it will become part of our daily lives.

18.1.1 Greenhouse effect (global warming)

We have by now all heard of the greenhouse effect, and most of us know that it is caused by trace gases in the atmosphere, which absorb and re-emit part of the infrared radiation that comes from the earth's surface. This leads to a warming of the lower atmosphere.

The greenhouse effect is not new; without it all the oceans would freeze. The main concern is the acceleration in the greenhouse effect due to increased levels of these trace gases, mainly CO_2, CFCs and methane. In the last 30 years, CO_2 alone has increased by 13% and is now increasing at 0.5% every year. CO_2 emissions are estimated to contribute 50% to the effect of global warming. Half of the CO_2 emitted in Western Europe results from the use of energy in buildings for heating, lighting and air conditioning.

18.1.2 Ozone depletion

Depletion of the ozone layer is mainly due to the use of CFCs, which react with ozone in sunlight to break down the layer. Full implications of this are not clear, but lead to increased ultraviolet penetration. This may cause increases in skin cancer, cataracts, reduced crop yields, damage to trees, even increased rates of degradation of building materials. About 10% of Europe's annual use of CFCs is related to buildings, mainly in refrigerants.

CFCs are estimated to contribute about 15% to global warming. Air conditioning accounts for about 50% of the demand for CFCs in refrigerants. They are thousands of times more harmful than CO_2, and their effects can be expressed as equivalent CO_2 emissions: for example, the greenhouse effect of 1 m³ of some CFCs is equivalent to between 1,000 and 25,000 m³ of CO_2.

18.1.3 What can those associated with construction do?

Clearly, any measures that reduce CO_2 emissions in constructing or running buildings are not just desirable, but essential. But would this be true whatever the capital cost? For example, solar energy systems (photovoltaic or PV systems) can convert the sun's energy directly to electricity, which can then be used in dwellings or commercial buildings. Their use would lead to significant reduction in energy consumption and CO_2 emissions. They are, however, as yet still very expensive and insufficiently reliable in service for mass use. Research continues and perhaps one day we may see PV systems as part of housing designs.

What is needed is a balance between environmental saving and capital cost – what we might call the **environmental equation**: the production of buildings, houses, factories and offices that are usable and flexible, requiring less energy to construct and consuming less energy in their daily use.

But above all they must be affordable: constructed at an economic cost that people are prepared to pay.

18.2 CONSTRUCTING BUILDINGS

In 1990, Gardiner & Theobald Research considered the energy that might be involved purely in the construction process. Concentrating on the structural frame alone, we examined the energy requirements to provide the frame in concrete, steel and timber, limiting the height of the buildings to four storeys and the gross area constructed to 10,000 m². The figures used assume manufacture of steel from scratch and ignore the fact that a good proportion of steel is produced from re-cycled material – using considerably less energy.

The results are as follows:

	Energy		CO_2 emission	
	Gigajoules	*kWh (000's)*	*t*	*m³*
Concrete	12,480	3,460	2,595	1,298
Steel	19,300	5,363	4,022	2,011
Timber	4,150	1,150	862	431

The relationship between kWh and CO_2 emissions is based upon conservative data sources. Other sources suggest higher CO_2 emissions: in some cases almost twice as much. It is worth noting that 1 kg of CO_2 occupies a volume of about 0.5 m³. The CO_2 released in producing a steel frame would equal about 50 times the volume of the finished building.

We compared these data with the annual energy requirements of our 10,000 m² of buildings, (a) when used as residential apartments and (b) when used as air-conditioned offices:

(a) Residential

	$t\ CO_2$ from 1 year's operation	$t\ CO_2$ released during construction of frame		
		Concrete	Steel	Timber
	929	2,595	4,022	862
% construction/ operation		279	432	93

(b) Air-conditioned offices

	$t\ CO_2$ from 1 year's operation	$t\ CO_2$ released during construction of frame		
		Concrete	Steel	Timber
	1,500	2,595	4,022	862
% construction/ operation		173	268	58

In our example the difference in CO_2 emission between the construction of steel and concrete frames represents the equivalent of that released by running a residential building for 1½ years. For a comparison between timber and steel frames the figure becomes 3½ years. It is the running of our buildings rather than the construction of our buildings that has the greatest impact on the environment.

However, our buildings all need structures. Careful, sensible choice of materials and methods of construction could produce environmental savings at little or no additional capital cost.

Can we perhaps look forward to a day when products are labelled to indicate the energy consumed in their production, much as foods are now labelled with energy content? Might we be able to select and specify products with low embodied energy?

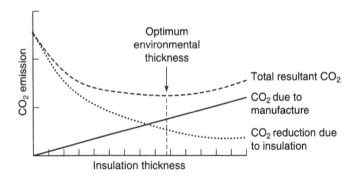

Fig. 18.1

Some work has already been undertaken in this field. Insulation materials are perhaps an easy illustration of this work.

18.2.1 Is more insulation an answer?

It is important when assessing energy saving measures to consider the CO_2 given off during manufacture and installation of the energy-saving mechanisms. The CO_2 given off in manufacture must not exceed the CO_2 emission that might be avoided during the life of the building.

Figure 18.1 illustrates how it is relatively easy to over-insulate and achieve no overall environmental benefit. The diagram is not specific, but rather illustrates the overall principles. Clearly, the optimum environmental thickness varies from material to material, although there will be a point where spending more money to increase insulation thickness in fact becomes more harmful to the environment. Again, perhaps continuing research will allow us to identify and label our insulating materials with their optimum environmental thickness, to assist future designers.

Research has already shown us that, for mineral fibre insulation 100 mm thick, the energy used in its manufacture is saved in about three years. Increase the thickness to 300 mm and this period becomes 25 years. For polystyrene foam the periods are roughly doubled. See Fig. 18.2.

18.2.2 An assessment method

In the UK we have no national energy plan. Our environmental legislation is embodied in various regulations. We do however lead Europe in having a

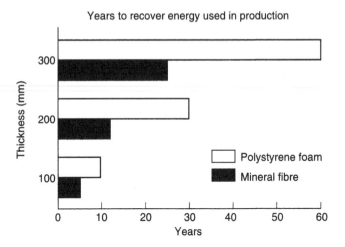

Fig. 18.2

comprehensive and widely used Environmental Assessment Method. Published by the Building Research Establishment, the guidelines are voluntary, and are aimed at encouraging designers, developers and users of buildings to become more environmentally sensitive.

What might be the effect of implementing such guidelines on say an air-conditioned office building in the City of London with the following criteria?

	Gross floor area	10,000 m²
	Capital cost	£10,000,000

Running costs		kWh/m² per annum	£
	Gas heating	215	26,600
30/W/m²	Office power	36	23,400
20/W/m²	Office lighting	50	32,500
	Landlord's power	3	1,950
	Landlord's lighting	3	1,950
	Air conditioning to 21°±1°C	43	27,950
			£114,350 per annum
Total CO$_2$ emission due to energy consumption			1,548 t per annum

A commentary on the BRE Assessment Guidelines and a detailed analysis of our findings is scheduled in the Appendix to this chapter. In brief, implementing all of the guidelines could add about 7% to the capital cost of the project, reduce annual running costs by about 12% and reduce annual CO_2 emissions by about 17%. It should, however be noted that while some of the issues have significant environmental benefit, others have considerable disbenefit.

18.3 LEANER/GREENER SPECIFICATION

It is now obvious that the major environmental savings come from implementing design guidelines that reflect global concerns. Could other reductions in standards improve the situation? Research in 1991/1992 has shown the evidence that the standards we were building to in the 1980s are excessive.

The provision of 30 W/m^2 of underfloor power was rarely being taken up. In many cases only 10 W/m^2 was being used. Why do we air condition to 21°C ± 1°C every day of the year?

What would the environmental impact be if we relaxed some of our commonly accepted criteria in favour of meeting occupiers' needs? Gardiner & Theobald Research have examined two alternative levels of specification, as follows.

Level 1
Reducing power and lighting loads and design air conditioning to 22°C ± 2½°C.

	Specification	*kW/m²/annum*
Gas heating		215
Reduce office power capacity	22 W/m²	27
Reduce office lighting capacity/standard	14 W/m²	35
Landlord's power		3
Landlord's lighting		3
New a/c design criteria	22°C±2.5°C	30
Total		313

Total CO_2 emissions due to energy consumption	1241 t/annum

Total running costs/annum   £90,300

Level 2

Further reducing power and lighting loads and designing air-conditioning to 3°C below outside ambient temperature, with a minimum of 21°C

	Specification	*kW/m²/ annum*
Gas heating		215
Reduce office power capacity	17W/m²	20
Reduce office lighting capacity/standard	10W/m²	25
Landlord's power		2
Landlord's lighting		2
New a/c design criteria	3° below ambient	20
Total		284

Total CO_2 emissions due to energy consumption   999 t/annum
Total running costs/annum   £71,450

Table 18.1 shows what effect the implementation of these two specification changes will have on our example building. Remembering that we included £150,000 for demolitions, which arguably is a normal cost for a city centre building and could perhaps be discounted, for an additional capital cost of perhaps only 3% we now have a building that provides less than half of the CO_2 per annum of our base building. Further reductions could of course be made if air-conditioning was not provided at all!

18.4 CONCLUSIONS

We have examined the impact on the environment of possible differences in construction of the frames of our buildings. This has shown us that the impact we have on the environment during construction is small when compared with the effects produced by running our buildings.

Our costed analysis of implementing one system of environmental guidelines suggests that we must pay a premium of about 7–8% on the cost of our buildings for 'going green': £775,000 to reduce CO_2 emissions by 260 t/yr, and running costs are only reduced by about £20,000/yr. This is

Table 18.1

	Capital cost	%	Running cost	%	CO_2 emissions	%
Base building	10,000,000	100.00	114,000	100.00	1548	100.00
BREEAM guidelines	775,000	7.75	(14,000)	(12.26)	(260)	(16.80)
	10,775,000	107.75	100,000	87.72	1288	83.20
Level 1 specification	(150,000)	(1.50)	(24,000)	(21.05)	(307)	(19.83)
	10,625,000	106.25	76,000	66.67	981	63.37
Level 2 specification	(180,000)	(1.80)	(19,000)	(16.67)	(242)	(15.63)
	10,445,000	104.45	57,000	50.00	739	47.74

assuming that we maintain our existing design criteria. It is only when we begin to relax the internal environmental criteria that we begin to see some real advantages: savings in capital cost, running cost and environmental impact, with more than half of the annual CO_2 output being removed.

It is clear that if we as users of buildings are to reduce the effect that we have on the environment, then we must learn a whole new set of rules by which to live and work. Occupiers must accept that they have a significantly greater impact on the environment than developers. Can we look forward to occupiers demanding 20 W/m^2 of power in lieu of 40 W/m^2? And here lies the real hub of the environmental equation. Capital cost injected by developers does not equal savings by occupiers. **One pays**, the other **benefits**.

Who should pay?

Perhaps what we need is a European environmental policy plan that transcends political and geographical divisions, becoming a sustained year-on-year programme, accepted as necessary and unavoidable by all, governments and citizens alike. Is it perhaps not reasonable for us all to expect to pay for maintaining our environment? Could we not all pay via European Community grants to buildings that achieve a higher than normal level of environmental rating?

We have to find a solution to the environmental equation: quite simply we cannot afford not to.

APPENDIX: COMMENTARY ON BRE ASSESSMENT METHOD GUIDELINES

It is worth noting that the BREEAM system operates on a credit or points basis, with one credit being issued for each of the design criteria met.

CFC emissions

For smaller developments, the use of R22 as a refrigerant has become acceptable. Bigger projects can necessitate the use of centrifugal chillers which, up until the Montreal Protocol, used CFC refrigerants. These are now banned, and substitute HCFCs such as R123 and HFCs such as R134a are being introduced.

The additional cost shown in our analysis of HCFCs or HFCs, with ODPs < 0.06 and < 0.03, is due to the additional equipment cost to replace traditional R11 machines. The running cost additions are primarily due to the lower efficiencies achievable with the new HCFCs and HFCs. The reduced efficiency results in higher electrical consumption, and for the 10,000 m^2 building, using an HCFC or HFC less than 0.03 ODP it is estimated that the reduced efficiency will be responsible for an additional 30 tonnes of CO_2 emissions per annum. This equals a volume of 15,000 m^3.

Halons

Halons for fire-extinguishing purposes are between three and ten times more damaging to the ozone layer than the already banned CFC refrigerants R11 and R12. They are still used in existing and new fire-extinguishing systems, particularly where CO_2 and H_2O alternatives are not viable.

Provided the halon is not discharged for testing purposes, the chances of emission of halon to the atmosphere is low, and consequently the overall environmental impact is not significant. However, if halon is not utilized, one credit is awarded.

Insulation materials

Credit is given to buildings using insulation that avoids the use of CFCs or HCFCs in its manufacture.

Insulating to higher standards in modern office buildings could be environmentally unfriendly. Taking our 10,000 m^2 office building, the balance point (that outside condition when heat loss equals heat gain) could be as low as 8 °C. If this is the case, mechanical cooling could be necessary for most of the time the building is occupied. Surely if we need to mechanically cool internal spaces even when the outside temperature is below 8 °C, we are not designing with the environment in mind. Should we be looking at fabrics that allow the heat out when internal gains exceed losses – particularly when external ambients are so low? Should there be less insulation in these circumstances?

Renewable timber sources

The project specification should state that timber and timber products come from managed and sustainable sources. In practice it is very difficult to ensure that all timber used complies with this requirement.

Recycling of materials

Space should be set aside for recycling materials.

Our building would need 10 m² of storage space for recycling paper in accordance with the guidelines.

Legionnaire's disease

Most modern office buildings are being designed with condensers or dry coolers, owing to the perceived difficulties associated with the maintenance and registration of wet cooling towers. This is unfortunate, because wet towers are less expensive to install and more efficient than dry coolers, although maintenance costs are higher.

The avoidance of *Legionnella pneumophila* is clearly essential, and credit is given either for properly designing a wet cooling tower system or for using air-cooled heat rejection, which does not act as a breeding ground for this bacterium. Unfortunately, the air-cooled solution involves additional energy consumption, which not only costs more money but produces more CO_2! This can be avoided if wet towers are correctly designed and maintained.

Local wind effects

Buildings should be designed with the local environment in mind. This in practice means buildings less than 20 m high or which satisfy an environmental wind assessment. Inevitably this will result in additional costs to the project without any obvious tangible return.

Re-use of existing site

We are encouraged towards the re-use of sites that have previously been built on or reclaimed from industrial processes or landfill. This is obviously an expensive option where the cost of demolition or specialist foundations may be necessary. Both options will necessitate the expenditure of energy, which might not be necessary on green field sites.

Domestic hot water systems

If CIBSE design procedures are adhered to, or point-of-use electric heaters are used, credit is awarded. Both in terms of cost and energy efficiency, point-of-use heaters are preferable and are commonly used in commercial developments.

Indoor air quality

Credit is given for designing fresh air volumes in accordance with CIBSE standards or for using controllable natural ventilation/openable windows. As our building is air-conditioned, the opening windows alternative is not applicable. Smoking can earn a credit if separately ventilated areas are designated for this activity. Because of the additional cost for a separate extract system it would be more economic to have a no smoking policy instead!

Humidity

Credit is awarded for the inclusion of a steam-based humidification system. This increases both cost and energy consumption significantly. While the cost aspects might be justifiable can the same be said for the introduction of an additional 40 tonnes of CO_2 into the atmosphere each year?

Hazardous materials

Credit is given for the avoidance of formaldehyde and exclusion of lead paint and asbestos. These materials have been omitted from buildings for several years and there is now no cost or environmental implication.

Lighting

The use of high-frequency ballasts instead of switch-start chokes for fluorescent lighting earns a credit. The cost penalty for installing these ballasts can be prohibitive but the energy saving, the better visual benefits plus the ability to dim the luminaires could be judged by some to be worthwhile. Under the present BREEAM scoring system no additional credit would be given.

BRE Assessment Method — effect on notional building

	Capital cost £	%	Running cost £/a	%	CO$_2$ emissions t/a	%
Base Building	10 000 000	100.00	114 000	100.00	1548	100.00
CFC emissions						
Ozone depletion less than 0.06	15 000	0.15	1000	0.88	13t	0.84
Ozone depletion less than 0.03	25 000	0.25	2000	1.75	26t	1.68
CFC leakage						
Leak detection	5000	0.05	–	0.00	–	0.00
Pump down container	5000	0.05	–	0.00	–	0.00
No halons	–		–	0.00	–	0.00
CFC free insulants	–		–	0.00	–	0.00
Renewable timber sources	–			0.00	–	0.00
Recycling material						
Addition of separate storage space	5000	0.05	3000	2.63	–	0.00
Cooling towers						
Wet to CIBSE Guidelines or	–		–			
air-cooled equipment	25 000	0.25	2000	1.76	26t	1.68
Wind effects						
Wind assessment tests	10 000	0.10	–	0.00	–	0.00
Models or > 20m high	–		–	0.00		
Re-use existing site						
Additional foundations or demolition may be required.						
Restricted access SAY	150 000	1.5	–	0.00	–	0.00
Water system						
Point of use water heaters	–	0.00	–	0.00	–	0.00
Carried forward	10 240 000	102.40	122 000	107.02	1613	104.20

BRE Assessment Method — effect on notional building

	Capital cost £	%	Running cost £/a	%	CO$_2$ emissions t/a	%
Brought forward	10 240 000	102.40	122 000	107.20	1613	104.20
Ventilation						
Ventilation to CIBSE guidelines or controllable natural ventilation	– N/a to a/c building	0.00	–	0.00	–	0.00
Separately ventilated areas for smokers	20 000	0.20		0.00	–	0.00
Humidity						
Steam based systems where humidification required	25 000	0.25	2000	1.75	26t	1.68
Hazardous materials						
Minimum release of formaldehyde	–		–		–	
No lead paint	–		–		–	
No asbestos	–		–		–	
Lighting						
High frequency ballasts	70 000	0.70	(3000)	(2.63)	(39)t	(2.52)
Sick Building Syndrome						
Daylight linking and use of task lights	30 000	0.30	(2000)	(1.75)	(26)	(1.68)
Meeting space and retreats	85 000	0.85	500	0.44	–	
Use indoor plants	45 000	0.45	2500	2.19	–	
Minimum energy						
Energy recovery equipment	100 000	1.00	(100 000)	(8.77)	(130)	(8.40)
Dimming/lighting controls	60 000	0.60	(4000)	(3.51)	(52)	(3.36)
BEMS	100 000	1.00	(8000)	(7.02)	(104)	(6.72)
	10 775 000	107.75	100 000	87.72	1288t/a	83.20

19

A European cost data bank

Douglas Robertson

19.1 INTRODUCTION

The term **construction economist** (CE) is used throughout this chapter to
embrace the activities of quantity surveyor, metreur-verificateur, and
geometro. However, it focuses on their role of obtaining value for money
for their clients – efficiency, effectiveness and economy – and the information they need.

The CE is a professionally qualified expert who advises clients on procurement routes, contractual arrangements, tendering markets, budgets and
costs-in-use. The CE provides clients with early cost advice, contract
documentation, financial control throughout the construction process, and
contractual and financial settlement with the contractor. These services are
applicable to new projects and refurbishment and take into account the
financial consequences of buildings in use. This role is performed as client
representative in design/build and turnkey contracts as well as open competitive tendering.

It is worth differentiating between the client/procurer role coupled with
quality performance control on the one hand and provider/contractor on the
other. The CE plays an important part within the contractor function but it
is different and not covered in this chapter.

Information and communications management is at the heart of the CE's
professional work and is especially important in an inherently disjointed
industry. The various players in the industry need information to be made
available at the right time, in a directly usable form, clear, reliable, consistent and unambiguous. Its purpose is to help decision-making. Information
is expensive, so it needs to be clearly necessary. Without sound information

the design and construction process would break down, and that would be even more expensive.

Computers have the ability to transfer data at the press of a button. The effort is in collecting, coding, classifying, verifying, entering and programming. Storage is not a problem and the equipment is no longer prohibitively expensive or bulky. Considerable effort has gone into the search for compatible systems and electronic data interchange (EDI). The work of Edifact and its standards and the applications of Edicon are showing the way ahead.

CAD is also a major influence, and the CE will be well advised to keep in touch with the structures that are becoming standards, which will dictate the development or replace some of the CE's techniques.

The profession has its roots in measurement, and quantification will continue to be a fundamental part of its work. It is much more than length × breadth × height, and embraces statistical methods of model building, sampling and forecasting as well as budgeting. It is however worth remembering that a recent piece of research revealed that measurements are taken on at least 15 different occasions during the design and construction process – not to mention the need for on-site measurement during a building's life. What value there would be in getting one measure to avoid subsequent measurements: CAD or geographical information systems could well revolutionize traditional thinking.

The CE has developed many skills and techniques that are common in principle but throughout the EC are different in practice and enshrined in different procedures. Most rely heavily on information to satisfy client requirements. CEs need to pool their information to strengthen the profession. There should be no hesitation for reasons of competition to withhold information as long as the following rules and criteria apply.

- The information provider must see value in collecting the data for their own use: they will not analyse it otherwise.
- The information provider must see value to them in being able to access a large data bank of useful information: they will not contribute otherwise.
- The data bank must be easily accessed and be of direct use by the CE to help him advise his client: subscribers will not pay to support it otherwise.
- The data must be trusted, relevant, consistent, comparable and up to date: subscribers will not rely on it otherwise.

- Subscribers who receive information from the data bank must also undertake to submit data to it: unity of purpose amongst CE subscribers will not be achieved otherwise.

So far this chapter has dwelt on the increasing dependence of the CE and other professionals in the building industry on information and its management. The rest of the chapter describes, with examples, how The Royal Institution of Chartered Surveyors has operated a CE's Building Cost Information Service (BCIS). The CEEC has asked BCIS to develop a service in Europe on its behalf. The vision is a centre of excellence for CE information: making a positive respectable start is the first objective.

19.2 BCIS

BCIS has operated since April 1962 as a collaborative venture for exchanging building cost information so that subscribers, involved in design and construction, can have ready access to the best available data related to construction economics. It is part of The Royal Institution of Chartered Surveyors, and is self-supporting from income from subscribers and from a number of research contracts. There are 1200 subscribers from professional firms, public service, commerce and industry.

BCIS operates as a reciprocal service exchanging tender price information mainly in the form of:

1. £/m² costs i.e.. cost analyses;
2. tender price indices.

This unique information is supported by other material relevant to construction economists:

3. construction statistics, e.g.. new orders, outputs etc.;
4. labour, hours and wages;
5. material costs;
6. cost indices;
7. analysis of market conditions;
8. daywork rates;
9. plant charges;
10. literature abstracts.

The data are also analysed by BCIS, which publishes various cost research studies including regular surveys of

11. keenness of contractors' pricing;
12. regional price variations;
13. contractor's mark-up;
14. ranges of £/m² by building type.

BCIS supplies data to its subscribers by:

- 14 bulletins in loose-leaf format – about 1,100 data sheets each year;
- a *Quarterly Review*, summarizing price trends;
- BCIS On-line – access to the computer host of all BCIS data;
- Occasional and specialist publications, e.g. *Standard Form of Cost Analysis, Guide to House Rebuilding Costs for Insurance Valuation, Dayworks Guide.*

The current development programme includes:

- establishing a network of regional analysts to interpret the BCIS regional tender price indices;
- expanding the information on mechanical and electrical engineering installations;
- creating a data bank of 'unit prices' for early cost advice on alternative design solutions and specifications.

BCIS has agreed to work with CEEC to establish the framework of a similar service for CEs in Europe. A number of BCIS services are described hereafter in some depth because of their relevance to establishing the CEEC data bank.

19.3 COST AND TENDER PRICE INDICES

It is necessary to differentiate between building costs and tender prices. The BCIS series of **building cost indices** measures the changes in costs of labour, materials and plant: i.e. the basic costs to the contractor. On the other hand, the BCIS series of **tender price indices** measures the trend of contractors' pricing levels in accepted tenders: i.e. the price that the client

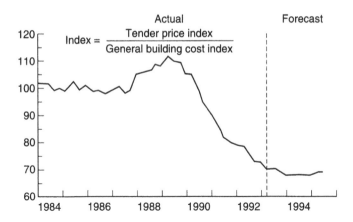

Fig. 19.1 Market conditions index.

has agreed to pay to the contractor to execute the work as defined in the contract documents.

BCIS also calculates an index of market conditions, which relates the trends in tender prices with the trends in building costs. It is a measure of keenness of competition in contractors pricing.

Figures 19.1–19.3 show the three indices (building costs, tender prices and market conditions, along with activity indicators of value of new work and value of output) since 1984. The graphs show the steady rise in costs and a markedly different trends in tender prices. From 1987 to mid-1990 tender prices rose faster than building costs, which is reflected in the peaking of market conditions and building activity. From the peak in 1989 tenders have become keener as the recession in the building industry continues.

Clients want to have some guidance on what costs and prices are going to do in the future so that they can judge when the best time will be to build and what provision they should make in their budgets for building inflation. BCIS produces a forecast looking 24 months ahead. A model is used to project costs and prices and the results are the basis of a discussion that simulates a contractor's 'tendering board' before a decision is finally made on the tender sum they are going to bid. BCIS states the assumptions used in the forecasts.

The BCIS Building Cost Indices are calculated from cost models of average buildings of different forms of construction, which were analysed to provide weighted work categories. Current costs of labour, materials and plant are then fed into the models. The BCIS series of cost indices includes the following:

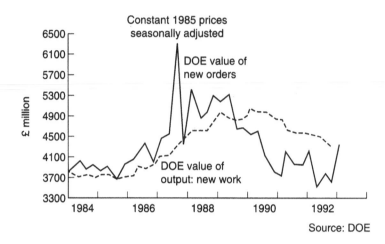

Fig. 19.2 Activity indicators.

- general building costs;
- steel-framed construction;
- concrete-framed construction;
- brick construction;
- mechanical and electrical engineering;
- basic labour;
- basic materials;
- basic plant.

The BCIS Tender Price Indices (TPI) are based on a random sample of accepted tenders for new building work with contract sums over £50,000, which have been priced in competition or by negotiation. The methodology used for the TPI was originally prepared for the Department of the Environment and it is used not just by BCIS but by the DoE, Scottish Development Department and Department of Health. In essence, bills of quantities are re-priced using a base schedule of rates, and the 'base' tender figure is compared with the actual tender figure to produce a 'project index'. The normal procedure is that items are selected to represent 25% of the value of the work in each section. The aim is to calculate 80 project indices each quarter to give a statistically reliable average result.

The BCIS TPI series includes the following:

- all-in TPI;
- housing TPI;

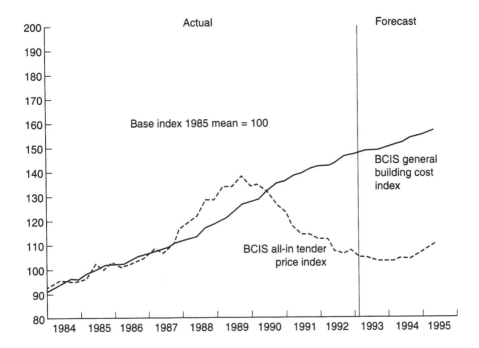

Fig. 19.3 Building cost trends.

- public sector TPI;
- housing TPI;
- private-sector TPI;
- private commercial TPI;
- private industrial TPI;
- refurbishment TPI.

Reliable series of building cost and tender price indices are fundamental data for the CE profession. To be able to produce an accurate measurement of these trends successfully obviously establishes the profession as the economist of the construction industry. The indices are fundamental tools of practice and are used to advise clients on:

- reporting on trends, market conditions and level of competition;
- forecasting trends and making budget projections;
- updating historical records, e.g. elemental analyses;
- approximate estimating and cost planning;

- linking contractual clauses to inflation;
- statistically analysing the competitiveness of tenders before acceptance;
- index linking for insurances.

19.4 ELEMENTAL ANALYSIS

The other main source of BCIS information is elemental cost analyses of accepted tenders. The purpose of cost analysis is to provide data that allow comparisons to be made between the cost of achieving various building functions in one project with that of achieving equivalent functions in other projects. This is the data bank for approximate estimating and cost planning.

Cost analysis allows for varying degrees of detail related to the design process; broad costs are needed during the initial period and progressively more detail is required as the design is developed. The elemental costs are related to square metre of gross internal floor area and also to a parameter more closely identifiable with the element's function, i.e. the element's unit quantity. More detailed analysis relates costs to form of construction within the element shown by 'all-in' unit rates. Supporting information on contract, design/shape and market factors is defined so that the costs analysed can be fully understood.

Table 19.1

Element of design criteria	Total cost of element (£)	Cost of element per m² of gross floor area (£)	Element unit quantity	Element unit rate (£)	Specification
3.B Floor finishes	13,262	17.29	694 m²	19.11	20 mm granolithic laid monolithic, no skirting
					2 mm thermoplastic tiles Series 2 on 48 mm cement and sand screed, softwood skirting
					2 mm vinylized tiles on 48 mm cement and sand screed, softwood skirting
					25 mm 'West African' sapele wood block floor on 40 mm cement and sand screed, softwood skirting
					12.5 mm red quarries on 32 mm screed, quarry skirting

			Floor finishes	(£)	Area (m²)	All-in unit rate (£)
			20 mm granolithic	185	30	6.17
			2 mm thermoplastic Series 2	184	13	14.15
Preliminaries 9.73% of remainder of contract sum			2 mm vinylized tiles	4717	395	11.94
			25 mm sapele blocks	5987	161	37.19
			12.5 mm quarries	2189	95	23.04

An example of a typical BCIS elemental cost analysis is shown in Fig. 19.4. BCIS publishes detailed cost analyses, concise cost analyses and £/m² costs of projects.

All-in unit rates are a more detailed stage in analysing costs, and show the costs of different specifications within one element. The example in Table 19.1 shows the all-in unit rates for various floor finishes.

The BCIS Standard Form of Cost Analysis was published in 1969, and includes the principles, instructions and definitions that are used throughout the CE profession in the UK. As a result of agreeing a standard format the profession has built up in BCIS and in their own offices a highly valued library of elemental cost information that satisfies the criteria of reliability.

The list of elements is critical, and as a step towards harmonization the members of CEEC have submitted their own lists of elements. A Concordance Document has been produced that allows CEEC member countries to interpret cost information and adjust the elemental allocations to their own standard. The following example from the Concordance shows the various ways the element **external walls** is dealt with.

UK	2E	external walls
Ireland	(21)	external walls
France	A211-2	loadbearing external walls or panels
	A231	infill to frames
		– infill masonry
		– curtain walls
		– cladding forming walls
	A221-3	parapets
	A235	projections
	A321-1	chimneys
Germany	31211	loadbearing external walls
	31311	non-loadbearing external walls
	31315	facades
	31316	external protective items (shutters, blinds)
Netherlands	(21)	external walls
Denmark	(21)	external walls
Spain	–	external walls, facades

BCIS On-line Analysis No. 11993

CI/SfB
320.
.   .   .   .

DETAILED COST ANALYSIS

Offices - 37 - a

BCIS Code: C - 4 - 5673

Job Title:	Transmitter Support Headquarters, Warwick Technology Park
Location:	Warwick, Warwickshire
Client:	British Broadcasting Corporation
Date for receipt: September 1987	Date of tender: September 1987

Indices used to adjust costs to 198
UK mean location base :-
TPI at tender 109; **TPI=1 mean 100
Location factor 0.99

INFORMATION ON TOTAL PROJECT

Project details:
4 storey building containing offices and technical workshops to support transmitter stations together with external works, services and drainage.

Site conditions:
Level green field site.

Market conditions:
Competitive. 3 stage negotiated tender, within budget. Design and build contract with design fee excluded from figures.

Tender documentation: Bill of approximate quantities

Selection of contractor: Negotiated

Number of tenders - issued: 1
                     received: 1

Type of contract: JCT Design and Build

Cost fluctuations: Fluctuating (no details)

Contract period - stipulated by client: -
                 - offered by builder: 17 months
                 - agreed: 17 months

Competitive Tender List

4,439,253

ANALYSIS OF SINGLE BUILDING

Accommodation and design features:
4 storey office and technical workshop block. Concrete Piled foundations and ground slab. Steel frame. Pitched timber trussed roof with tiles; concrete paved flat roof. Facings/block cavity walls. Aluminium windows; automatic entrance doors. Brick and block internal walls; Formica and demountable partitions. Hardwood internal doors. Plaster, tiles and wallpaper to walls; anti static carpet tiles to floors; plasterboard and suspended ceilings. Fittings. Oil and gas fired heating. Sanitary, ventilatio and electrics. Lift. Fire and burglar alarms. PA.

Areas:

		Functional units:
Basement floors	-	3,759 m2 usable floor area
Ground floor	2,290 m2	
Upper floors	3,383 m2	
Gross floor area	5,673 m2	Percentage of gross floor area
Usable area	3,759 m2	
Circulation area	1,120 m2	
Ancillary area	648 m2	
Internal divisions	146 m2	
Gross floor area	5,673 m2	

		Storey Heights:		
Floor space not enclosed.	-		Average	
Internal cube	21,047 m3	below ground floor	-	
External wall area	3,127 m2	at ground floor	3.80 m	
Wall to floor ratio	0.55	above ground floor	3.53 m	

BRIEF COST INFORMATION

TOTAL CONTRACT

Measured work	3,943,378	
Provisional sums	48,000	
Prime cost sums	25,000	
Preliminaries	422,875	- being 10.53 % of remainder of
Contingencies	-	contract sum (less contingencies)
Contract sum	4,439,253	

Functional unit cost excluding external works - at tender date   985.42 per m2 usable floor area
                                              - at 1985, UK mean location   913.19 per m2 usable floor area

Fig. 19.4

:I/SfB
320.

Offices - 37 - b

ELEMENT COSTS

Gross internal floor area: 5,673 m2   Date of tender: September 1987

lement		Preliminaries shown separately				Preliminaries apportioned		
		Total cost of element	Cost per m2 gross floor area	Element unit quantity	Element unit rate	Total cost of element	Cost per m2 gross floor area	Cost per m2 at 1985,UK mean location
	SUBSTRUCTURE	299,593	52.81			331,136	58.37	54.09
A	Frame	304,567	53.69			336,634	59.34	
B	Upper floors	78,260	13.80			86,500	15.25	
C	Roof	181,806	32.05			200,948	35.42	
D	Stairs	48,980	8.63			54,137	9.54	
E	External walls	404,291	71.27			446,858	78.77	
F	Windows & external doors	250,364	44.13			276,724	48.78	
G	Internal walls and partitions	52,209	9.20			57,706	10.17	
H	Internal doors	49,245	8.68			54,430	9.59	
	SUPERSTRUCTURE	1,369,722	241.45			1,513,937	266.87	247.30
A	Wall finishes	301,136	53.08			332,842	58.67	
B	Floor finishes	41,928	7.39			46,343	8.17	
C	Ceiling finishes	77,561	13.67			85,727	15.11	
	INTERNAL FINISHES	420,625	74.15			464,912	81.95	75.94
	FITTINGS	279,891	49.34			309,360	54.53	50.53
A	Sanitary appliances	21,346	3.76			23,594	4.16	
B	Services equipment	-	-			-	-	
C	Disposal installations	-	-			-	-	
D	Water installations	33,775	5.95			37,331	6.58	
E	Heat source	104,664	18.45			115,684	20.39	
F	Space heating & air treatment	179,329	31.61			198,210	34.94	
G	Ventilating systems	101,121	17.82			111,768	19.70	
H	Electrical installations	340,669	60.05			376,537	66.37	
I	Gas installations	-	-			-	-	
J	Lift & conveyer installations	50,865	8.97			56,220	9.91	
K	Protective installations	98,336	17.33			108,690	19.16	
L	Communications installations	15,598	2.75			17,240	3.04	
M	Special installations	11,042	1.95			12,205	2.15	
N	Builder's work in connection	24,746	4.36			27,351	4.82	
O	Builder's profit & attendance	-	-			-	-	
	SERVICES	981,491	173.01			1,084,830	191.23	177.21
	BUILDING SUB-TOTAL	3,351,322	590.75			3,704,175	652.95	605.08
5A	Site works	444,656	78.38			491,473	86.63	
5B	Drainage	186,498	32.87			206,134	36.34	
5C	External services	33,902	5.98			37,471	6.61	
5D	Minor building works	-	-			-	-	
	EXTERNAL WORKS	665,056	117.23			735,078	129.57	120.07
	PRELIMINARIES	422,875	74.54			-	-	
	TOTAL (less contingencies)	4,439,253	782.52			4,439,253	782.52	725.16

Fig. 19.4 (continued)

CI/SfB
320.
.   .   .   .   .

SPECIFICATION AND DESIGN NOTES

Offices - 37 - c

1		SUBSTRUCTURE	150mm piles, concrete Grade C40X in reinforced pile caps, 250mm floor and 300mm foundations.
2A	Frame		Structural steel frame.
2B	Upper floors		Precast flooring by specialist sub-contractor.
2C	Roof		Pitched timber trussed roof with interlocking tiles; concrete paved flat roof. PVC rainwater goods.
2D	Stairs		Mild steel with stainless steel balustrades; mild steel with PVC cvered handrail. External fire escpaes.
2E	External walls		Buff facings and concrete block cavity walls.
2F	Windows & external doors		Aluminium windows. Automatic entrance doors.
2G	Internal walls and partitions		Brick and block internal walls; some Formica cubicles and demountable Neslo partitions.
2H	Internal doors		Fire resist hardwood doors.
3A	Wall finishes		Plaster and either emulsion or Muraspec vinyl wallpaper. Langley Buchtal wall tiles.
3B	Floor finishes		Anti-static carpet tiles.
3C	Ceiling finishes		Plasterboard and skim; Armstrong Second Look 1 suspended ceilings.
4		FITTINGS	Catering equipment, bar fittings, window blinds, reception desk, TV supports, partitions and acoustic and fire barriers.
5A	Sanitary appliances		Sanitary appliances.
5D	Water installations		Water installations.
5E	Heat source		Oil and gas fired burners with water treatment plant and ventilation and comfort cooling sytem (centralised) to all parts of the building.
5H	Electrical installations		Electrical installations.
5J	Lift & conveyer installations		1000kg (13 person) lift.
5K	Protective installations		Contractor designed.
5L	Communications installations		Alarm, fire alarm, PA systems.
5M	Special installations		Kitchen equipment, special water for laboratories etc.
5N	Builder's work in connection		Normal builder's work in connection with services.
6A	Site works		Paving, roadworks and landscaping.
6B	Drainage		Drainage installation.
6C	External services		Water, gas, electric and telephone services.
7		PRELIMINARIES	10.53% of remainder of Contract Sum (excluding Contingencies).

CREDITS

SUBMITTED BY	Barker, Barton & Lawson
CLIENT	British Broadcasting Corporation
ARCHITECT	Tarmac Construction Ltd
QUANTITY SURVEYOR	Barker, Barton & Lawson
CONSULTING ENGINEERS	Ove Arup & Partners
GENERAL CONTRACTORS	Tarmac Construction Ltd

Fig. 19.4 (continued)

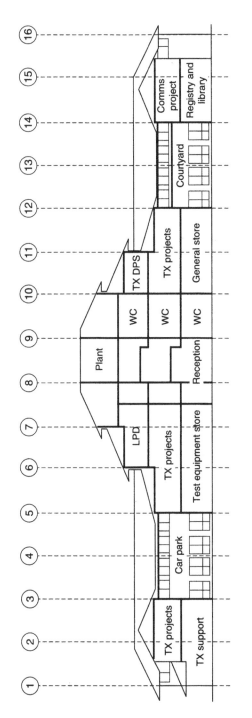

Fig. 19.4 (continued)

The use of elements has other applications as well as for cost analysis: indeed, a recent ISO report identified 14 uses for elemental classification. Element lists are seen by classificationists as an integral part of a standard classification system for the construction industries. There is currently a window of opportunity that, if taken now, could in time lead to an internationally agreed list of elements. This worthy cause was advanced at a recent forum in Washington DC on 'International Classification of Construction Elements' where BCIS took a leading role in support of CEEC objectives. Out of this meeting the following definition of an element emerged as acceptable to all participants:

> The physical parts and systems of a building or other facility each with a characteristic function. Elements are defined without regard to the type of technical solution or the method or form of construction.

A programme of further work was agreed, and BCIS is charged with leading the work on building elements, so CEs will be strongly represented in these developments.

The BCIS definitions of three parameters used in cost analysis are also worth recording:

- **Gross internal floor area**: total of all enclosed spaces fulfilling the functional requirements of the building measured to the internal structural face of the enclosing walls. It includes areas occupied by partitions, columns, chimney breasts, internal structural or party walls, stairwells, lift wells and the like. Also included are lift, plant, tank rooms and the like above main roof slab. Sloping surfaces such as staircases, galleries, tiered terraces and the like are measured flat on plan.
- **Element unit quantity**: a suitable unit that relates solely to the quantity of the element itself; e.g. in the case of 'Wall finishes' the element unit quantity would be 'Total area of the wall finishes in m^2' and in the case of 'Heat source' the element unit quantity would be 'kilowatts'.
- **Functional unit**: the number of units of accommodation, e.g. seats in churches, school places, hospital ward beds, or the net usable floor area, e.g. offices, factories.

Elemental cost analyses are kept by most CEs. As with most information, each CE will rely on his/her own records because of their familiarity and intimate understanding of the circumstances of the project. However, the

BCIS experience is that CEs value having access to a large data bank of cost analysis and are willing to contribute to it.

CEs use elemental cost analyses as the source of information for:

- early cost advice;
- approximate estimating;
- cost planning and cost control;
- comparing the cost of alternative designs and specifications.

19.5 BCIS ON-LINE AND APPROXIMATE ESTIMATING SOFTWARE

BCIS On-Line provides access to its computer-held cost data banks 24 hours a day, 365 days a year. An On-line subscriber needs a PC, a modem and a telephone line. BCIS provides the software to connect to the Micro VAX 3100–80 and to select and download the information onto the subscriber's own microcomputer.

On-Line was introduced in 1984 and has been developing ever since then. The computer host contains all BCIS data and is continuously being updated to give subscribers current information. Figure 19.5 shows what information is held on the data bank (e.g. national insurance rates, CITB levy, dayworks, cost analyses, indices etc.) and how it can be accessed, selected, downloaded, and printed.

BCIS has written sophisticated software to allow On-Line subscribers to manipulate the cost information on their own microcomputer without staying on-line with BCIS VAX. The BCIS Approximate Estimating Package (AEP) uses elemental cost analyses to calculate a budget estimate and cost plan for a new project. The AEP is fully integrated with the rest of the On-Line service and can also use subscriber's own data sources. All BCIS software is accompanied by full documentation and user guides.

19.6 BUILDING MAINTENANCE INFORMATION (BMI)

The CE has to take a long view when he offers cost advice. Life-cycle techniques are used to access alternative solutions by taking into account not only initial costs but also running and occupation costs. Property occupancy cost planning is a move to bring planned and costed programmes of work higher on the CE's and client's agenda. Maintenance audit and facilities management widen the subject further.

In the UK, building maintenance alone amounted to £26,120,000,000 in 1991. It is 43% of total construction output. Each year maintenance expenditure is equivalent to only 1.5% of the gross capital stock of buildings and works. Add to that cleaning, energy, security, porterage etc. and it becomes clear why The Royal Institution of Chartered Surveyors runs a sister organization to BCIS called BMI.

The same formula applies, but BMI receives less information from property owner/occupiers than BCIS does from its subscribers. More reliance therefore has had to be placed on synthetic data than on analysis.

The main publications of BMI are:

- cost indices for general maintenance, broken down by sectors;
- cost indices for redecorations, fabric maintenance, M&E maintenance, lift maintenance, cleaning, energy (coal, gas, oil, electricity);
- earnings and material prices;
- property occupancy cost analyses (BMI publishes the *Standard Form of Occupancy Cost Analysis*)
- *BMI Price Book* (containing labour constants, material prices and prices of maintenance items).

19.7 CEEC MARKET RESEARCH

CEEC is currently conducting two pieces of market research amongst its executive members. The first aims to collect personal profiles of CEs throughout Europe: 'Who is Who in Construction Economics'. The second part of this approach aims to identify what information is available in each country, what information is used and valued by CEs and what information they would be willing to exchange.

It would be disappointing and surprising if CEEC were not able to show professional collaboration and a willingness to exchange information, share knowledge and develop good practice.

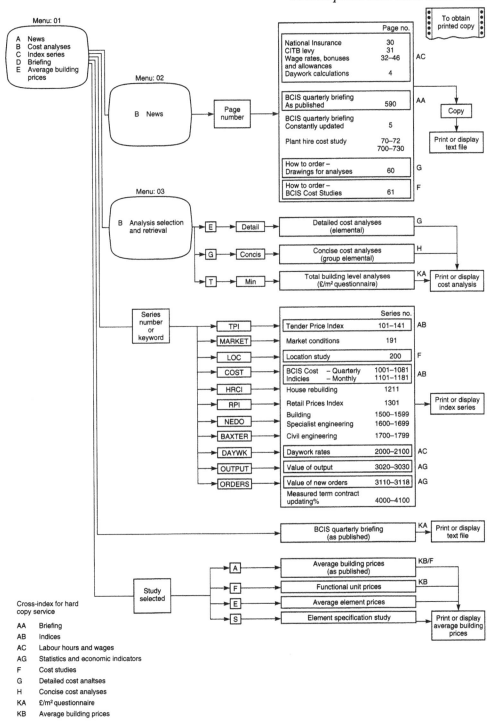

Fig. 19.5 Up-to-date data available on BCIS On-Line.

Index